PHYSICS
OF THIN FILMS

# PHYSICS
# OF THIN FILMS

By Ludmila Eckertová

PLENUM PRESS ● NEW YORK AND LONDON

SNTL ● PUBLISHERS OF TECHNICAL LITERATURE, PRAGUE

Distributed throughout the world with the exception of the Socialist countries by

Plenum Press,
a Division of Plenum Publishing Corporation, 227 West 17th Street, New York 10011
ISBN 0-306-30910-6
Library of Congress Catalog Card Number 76-12177

© 1977 Ludmila Eckertová
Translated by Pavel Bratinka
English edition first published in 1977 simultaneously by
Plenum Publishing Corporation and
SNTL - Publishers of Technical Literature, Prague

Printed in Czechoslovakia

CONTENTS

1. Introduction

2. Methods of Preparation of Thin Films

2.1 Chemical and Electrochemical Methods . . . . . . . . . . . .  14
2.2 Cathode Sputtering . . . . . . . . . . . . . . . . . . . . .  18
2.2.1 Principle of Diode Sputtering . . . . . . . . . . . . . . .  18
2.2.2 Some Special Systems of Cathode Sputtering . . . . . . . . .  23
2.2.3 Low-Pressure Methods of Cathode Sputtering . . . . . . . . .  26
2.3 Vacuum Evaporation . . . . . . . . . . . . . . . . . . . . .  28
2.3.1 Physical Foundations . . . . . . . . . . . . . . . . . . .  28
2.3.2 Experimental Techniques . . . . . . . . . . . . . . . . . .  33
2.3.21 Evaporation Apparatus . . . . . . . . . . . . . . . . . .  33
2.3.22 Substrates and Their Preparation . . . . . . . . . . . . .  38
2.3.23 The Most Important Materials for Evaporation . . . . . . . .  42
2.3.24 Evaporation Sources . . . . . . . . . . . . . . . . . . .  45
2.3.25 Special Evaporation Techniques . . . . . . . . . . . . . .  46
2.3.26 Masking Techniques . . . . . . . . . . . . . . . . . . . .  49

3. Thin Film Thickness and Deposition Rate Measurement
Methods

3.1 Balance Methods . . . . . . . . . . . . . . . . . . . . . .  53
3.1.1 Microbalance Method . . . . . . . . . . . . . . . . . . . .  53
3.1.2 Vibrating Quartz Method . . . . . . . . . . . . . . . . . .  54
3.2 Electrical Methods . . . . . . . . . . . . . . . . . . . . .  57
3.2.1 Electric Resistivity Measurement . . . . . . . . . . . . . .  57
3.2.2 Measurement of Capacitance . . . . . . . . . . . . . . . . .  57
3.2.3 Measurement of Q-factor Change . . . . . . . . . . . . . . .  58
3.2.4 Ionization Methods . . . . . . . . . . . . . . . . . . . . .  59
3.3 Optical Methods . . . . . . . . . . . . . . . . . . . . . .  60
3.3.1 Method Based on Measurements of Light Absorption Coefficient .  60
3.3.2 Interference Methods . . . . . . . . . . . . . . . . . . . .  61
3.3.3 Polarimetric (Ellipsometric) Method . . . . . . . . . . . .  67
3.4 Deposition Rate Monitoring Using Transfer of Momentum . . . .  68
3.5 Special Thickness Monitoring Methods . . . . . . . . . . . .  69
3.5.1 Stylus Method . . . . . . . . . . . . . . . . . . . . . . .  69

3.5.2 Radiation-absorption and Radiation-emission Methods . . . . . 70
3.5.3 Work-function Change Method . . . . . . . . . . . . . 71

## 4. Mechanism of Film Formation

4.1 Formation Stages of Thin Films . . . . . . . . . . . . . 72
4.2 Nucleation . . . . . . . . . . . . . . . . . . . . . . . 73
4.2.1 Capillarity Theory of Nucleation . . . . . . . . . . . . . 76
4.2.2 Statistical (Atomistic) Theory of Nucleation . . . . . . . . . . 81
4.2.3 Influence of Individual Factors on Nucleation Process . . . . . . 86
4.2.4 Some Experiments for Verification of Nucleation Theories . . . . 91
4.3 Growth and Coalescence of Islands . . . . . . . . . . . . 93
4.4 Influence of Various Factors on Final Structure of Film . . . . . 97
4.4.1 Special Properties of Films Deposited by Cathode Sputtering . . 98
4.5 Crystallographic Structure of Thin Films . . . . . . . . . . . 102
4.6 Epitaxial Films . . . . . . . . . . . . . . . . . . . . 110

## 5. Composition, Morphology and Structure of Thin Films

5.1 Methods for Determination of Chemical Composition of Films . . 115
5.2 Electron Microscopy of Thin Films . . . . . . . . . . . . 118
5.2.1 Transmission Electron Microscopy . . . . . . . . . . . . 118
5.2.2 Electron-microscopic Examination of Surface by Replica Method . 128
5.2.3 Special Types of Electron Microscopes for Direct Image-forming of
  Film Surface . . . . . . . . . . . . . . . . . . . . . 130
5.2.31 Scanning Microscope . . . . . . . . . . . . . . . . . 130
5.2.32 Reflection Microscope . . . . . . . . . . . . . . . . . 131
5.2.33 Emission Microscopes . . . . . . . . . . . . . . . . . 132
5.2.4 Tunnel Emission and Field Ionization . . . . . . . . . . . 133
5.2.41 Field Electron Microscope . . . . . . . . . . . . . . . 135
5.2.42 Field Ion Microscope . . . . . . . . . . . . . . . . . 138
5.3 Diffraction of Electrons . . . . . . . . . . . . . . . . . 139
5.3.1 Diffraction of High-Energy Electrons in Transmission and in
  Reflection . . . . . . . . . . . . . . . . . . . . . . 140
5.3.2 Low-Energy Electron Diffraction (LEED) . . . . . . . . . . 144
5.4 X-ray Methods . . . . . . . . . . . . . . . . . . . . . 147
5.4.1 X-ray Diffraction . . . . . . . . . . . . . . . . . . . 147
5.4.2 X-ray Microscopy . . . . . . . . . . . . . . . . . . . 147
5.5 Auger Spectroscopy . . . . . . . . . . . . . . . . . . . 148

## 6. Properties of Thin Films

6.1 Mechanical Properties . . . . . . . . . . . . . . . . . . 151
6.1.1 Experimental Methods for Measurement of Mechanical Properties
  of Thin Films . . . . . . . . . . . . . . . . . . . . . 152
6.1.2 Stress in Thin Films . . . . . . . . . . . . . . . . . . 155
6.1.3 Mechanical Constants of Thin Films . . . . . . . . . . . . 157
6.1.4 Adhesion of Thin Films . . . . . . . . . . . . . . . . . 158
6.1.5 Rayleigh Surface Waves . . . . . . . . . . . . . . . . . 160

6.2 Electrical and Magnetic Properties of Thin Films . . . . . . . . 162
6.2.1 Conductivity of Continuous Metal Films . . . . . . . . . . . 163
6.2.2 Conductivity of Discontinuous Metal Films . . . . . . . . . 170
6.2.3 Electrical Properties of Semiconducting Thin Films . . . . . . 176
6.2.4 Galvanomagnetic Effects in Thin Films . . . . . . . . . . . 180
6.2.5 Superconductivity in Thin Films . . . . . . . . . . . . . 184
6.2.6 Conductivity of Thin Dielectric Films . . . . . . . . . . . 196
6.2.7 Dielectric Properties of Thin Films . . . . . . . . . . . . 209
6.2.8 Ferromagnetic Properties of Thin Films . . . . . . . . . . 213
6.3 Optical Properties of Thin Films . . . . . . . . . . . . . 221

# 7. Application of This Films

7.1 Optical Applications . . . . . . . . . . . . . . . . . . 224
7.2 Applications in Electronics . . . . . . . . . . . . . . . . 227
7.2.1 Electric Contacts, Connections and Resistors . . . . . . . . 228
7.2.2 Capacitors and Inductances . . . . . . . . . . . . . . . 230
7.2.3 Applications of Ferromagnetic and Superconducting Films . . . . 231
7.2.4 Active Electronic Elements . . . . . . . . . . . . . . . 236
7.2.5 Microacoustic Elements Using Surface Waves . . . . . . . . 238
7.2.6 Integrated Circuits (IC) . . . . . . . . . . . . . . . . 240
7.2.7 Thin Films in Optoelectronics and Integrated Optics . . . . . . 242
7.2.8 Further Applications . . . . . . . . . . . . . . . . . . 245
References . . . . . . . . . . . . . . . . . . . . . . . 246
Index . . . . . . . . . . . . . . . . . . . . . . . . . 248

# CHAPTER 1

# INTRODUCTION

The investigation of the physical properties of matter has progressed so much during the last hundred years that today physics is divided into a large group of special branches, which are often very distant from each other. These branches arise because of the vast extent of the science itself, and are distinguished by the particular area studied, the method of investigation and so on. An independent and important branch that has developed recently is the physics of thin films. This deals with systems which have only one common property, namely, that one of their dimensions is very small, though all other physical properties of such systems may be different, as well as methods of investigating them.

Usually, we investigate the physical characteristics of three-dimensional bodies. Their characteristic properties are often related to a unit volume, i.e. it is assumed that they are volume-independent. This assumption is legitimate as long as the dimensions are 'normal', i.e. more or less within macroscopic limits; but as soon as one dimension becomes so small that there is a considerable increase in a surface-to-volume ratio, that assumption is no longer valid.

In the bulk volume of a body, forces act upon a given particle (atom, electron). In crystals these forces are of a periodic nature; in amorphous substances, in which there is at most only a short-range order, they have no such periodicity; but in both cases the particles are under the influence of forces from all directions. There is, however, a cut off of such forces when the surface area only is considered. The forces acting upon the particles at the surface are different from those of the bulk, the main difference being their pronounced asymmetry. Energy states at the surface may thus be substantially different from internal ones, and we speak, therefore, about the existence of surface states.

If we consider a very thin film of some substance, we have a situation in which the two surfaces are so close to each other that they can have a decisive influence on the internal physical properties and processes of the substance, which differ, therefore, in a profound way from those of a bulk material. The decrease in distance between the surfaces, and their mutual interaction, can result in the rise of completely new phenomena. Further, the reduction of one dimension of a material to an order of only several atomic layers creates an intermediate system between macro-systems and molecular systems, thus providing us with a method of investigation of the microphysical nature of various processes.

These are some of the reasons why thin films have attracted the attention of physicists, and why a whole branch of physics devoted to thin films has been created, and why the related technological branches have developed.

It is not possible to answer specifically the question, what is the limit within which a film should be considered 'thin'. It is possible to say in general that the limit is determined by the thickness under which the described anomalies appear, but this differs for different physical phenomena. In practice, the physics and technology of TF deals with films of between tenths of a nanometre and several micrometres.

The most conspicuous phenomena associated with TF are optical ones, especially that of interference colors, which can be commonly observed, e.g. on the thin film of oil spilled on water or on a wet pavement. These phenomena attracted the attention of physicists as early as the second half of the seventeenth century, their discovery and explanation being associated with the names of Boyle, Hooke and Newton. Two hundred years later the optics of TF were advanced by the measurements of Jamin, Fizeau, Quincke and others and the theoretical works of Drude. TF interference provided the means for exact measurement of TF thickness and found applications in optical and other fields. The application of anti-reflection filters and of various decorative coatings utilizing the interference colors, e.g. in bijoux production, are widely known today.

The preparation of TF by vacuum deposition methods, i.e. by cathode sputtering and vacuum evaporation, is another subdivision of TF physics that had its origin in the mid-nineteenth century, but did not undergo substantial development until relatively recently. Both phenomena were observed during the study of other physical processes (sputtering in the study of gas discharge by Grove, evaporation in the study of light sources with carbon filaments, by, for example, Edison and Fleming).

To the same period belong the work of Jamin and Magnus on the condensation of vapors and gases on the surface of solid substances, demonstrating the importance of adsorbed films and the connection with chemical surface processes — such as catalysis — and also the work of Faraday and Young, who investigated films prepared by chemical and electrochemical methods. The rectifying effect of an aluminium electrode covered by a thin film of the oxide was discovered (Pollak, Graetz), pointing to the possibility of the use of TF in electronics. Beetz used TF elements for the investigation of magnetic phenomena, opening up a very promising road from both the theoretical and practical point of view.

The study of the surface tension of liquids has developed into the investigation of monomolecular films of organic substances on the surface of liquids; this is important in the investigation of various biochemical and physiological processes. In addition, the study of surface forces in TF of organic substances plays a role in friction mechanics. Thin surface films on various materials, especially metals, both natural (e.g. oxides or other compounds derived from substrate material) and artificially created or deposited, have long been the subject of research by technologists interested in problems of corrosion and protection of materials.

From the beginning of this century the electric properties of thin films have been studied, from measurement of conductivity to the study of superconductivity, as well as the emission of electrons from TF. This research has made extraordinarily rapid advances in recent years.

The vigorous development of electronics during and since the second world war has brought about a steady decrease in the dimensions of electronic equipment. This has been stimulated by space exploration with its demand for intricate radioelectronic equipment of high reliability, small size and weight. Recently, too, there has been considerable development in the field of medical electronics. This branch of electronics requires electronic instruments which can be placed on body surfaces of animal or man, or, if need be, implanted to measure or stimulate and control various vital processes.

A further stimulation for the development of miniaturized electronic equipment was the development of computers, which while becoming more and more complicated need a maximum reliability and the smallest possible dimensions of the elements.

This miniaturizing of classical electronic elements (tubes, resistors, capacitors) was followed by the use of semiconductor elements, e.g. diodes, transistors, etc., on printed wiring. A further step towards microminiaturization was the introduction of micromodules, small ceramic plates (e.g. $20 \times 20$ mm), on which the passive elements are prefabricated, mostly in the form of thin films (resistances and capacitors), from which it is possible to build functional units in a very compact form.

The mastery of the technology of producing active semiconductor elements, i.e. transistors made from monocrystals of germanium and silicon, has led to the production of integrated circuits. These are produced in two ways. In the first, separate components are made by a common semiconductor technology and these are then interconnected by evaporation. In the second, all components are made in the crystal and interconnected by the material of the crystal with suitably modified conductivity. On the surface of the crystal, there are only the output and input of the system, which can be a two-stage amplifier, a flip-flop circuit, etc., or if necessary, a whole series of such elementary circuits (so-called monolithic integrated circuits).

New possibilities of microminiaturization have been opened up by the use of TF not only to connect the individual elements, but also as the elements themselves, both active and passive. This application exploits the fact that one of the dimensions is almost zero (from the macroscopic point of view) and that the thickness of the element is determined only by the thickness of the substrate on which the TF is deposited.

Up to the present, only the problem of passive elements has been solved satisfactorily and it has been necessary to overcome a number of difficulties with the reproducibility and stability of their characteristics. In the domain of active elements (i.e. the elements replacing tubes and transistors) research has not gone far beyond the preliminary stage and new principles of TF transistor operation are being sought. Hybrid circuits, combining TF passive and active elements made by semiconductor technology, are being successfully produced. There already exist purely TF integrated circuits with field-effect transistors. We shall encounter these applications again in the final chapter.

The rapid development of TF technology, especially that of vacuum-deposited TF — i.e. evaporation and cathode sputtering — and the exacting requirements of stability and reproducibility in electronic applications have stimulated the development of basic research in TF physics. It was realized that the demands of the industry could not be reliably met if knowledge of the basic laws of TF formation was inadequate, and if their relationship to the various physical parameters of processes involved in production and to their mechanical and electrical properties were not understood.

Detailed investigation of the structure of TF and processes involved in their formation has been made possible by two physical methods, namely, electron microscopy and electron diffraction. Electron microscopy enables us not only to study the morphology of the films, but also to observe the process of formation of film by evaporation directly in the viewing field. As the resolving power of the best electron microscopes reaches the value 0·5 nm, we can see that this instrument can provide valuable information. The diffraction of electrons due to the wavelike nature of electrons passing through a crystal lattice was discovered by Davisson and Germer in 1927. The electron waves cancel or reinforce each other, depending on the direction of propagation, in such a way that after the impact of electrons upon a screen or photographic plate they give rise to a number of light (black in the case of the plate) spots. It is possible to discover from their positions and intensities whether a substance is amorphous, polycrystalline, or monocrystalline and, if necessary, what kind of lattice it has and how it is orientated. A special case of diffraction is low-energy electron diffraction (LEED). The basis of both phenomena is identical, the only difference being that low-energy electrons only penetrate to a depth of several atomic layers and therefore yield information only about the state and structure of an extremely thin layer at the surface of the sample and, accordingly, are suitable for the study of surface processes and properties.

A further condition which made possible a considerable development of both thin film physics and its application during the last ten years or so was the development of vacuum physics and technology. It was discovered that for some purposes, especially in basic research and in a number of applications, it is necessary to prepare thin films in extremely clean conditions in which it is possible to maintain the surface of the substance without adsorption for sufficiently long periods. When we take into account the fact that in the pressures commonly used in high-vacuum apparatus, i.e. pressures of the order of $10^{-6}$ torr, the monomolecular adsorbed layer arises on the clean surface after $\sim 1$ second, it is clear that satisfactory results can be achieved only by the use of ultra-high vacuum equipment which is able both to provide and measure pressure in the region of $10^{-9}$ torr and lower.

With the help of these modern techniques, TF physics has made considerable progress during recent years and the speed of its development is accelerating in keeping with the overall tendency of the development of science in our society. The number of papers from this branch (see e.g.[33]) has become an avalanche and even for the specialist it is impossible to keep them all under review.

An author of a work on TF therefore faces an almost insurmountable problem in writing a book that will not become obsolete before it gets into the hands of the reader. Hence the present writer apologizes in advance if by the time of publication some statements in this book have already been overridden by the latest research.

CHAPTER 2

# METHODS OF PREPARATION
# OF THIN FILMS

Methods of preparing thin films may be divided essentially into two main groups, namely chemical methods (including electrochemical methods) and physical methods. We shall devote most attention to physical methods, because they result in the formation of very pure and well-defined films. In practice they are applicable to all substances and to a great range of thicknesses. But first in the section that follows we shall briefly enumerate the chemical methods that may be used in producing thin films.

## 2.1 Chemical and Electrochemical Methods

Among chemical and electrochemical methods the most important are electrolytic deposition, electroless deposition, anodic oxidation and chemical vapor deposition.

(*a*) In cathode electrolytic deposition, the substance (metal) to be deposited is present in a solution or melt in the form of ions. If we insert two electrodes into the solution (or melt) the positive ions of the metal will be attracted to the cathode where the metal will be deposited. The mass of the substance deposited is proportional to the amount of electricity used. The proportionality constant is the electrochemical equivalent of the given substance, which, for example, is $2.04 \cdot 10^{-3}$ g/C for Au and $0.09 \cdot 10^{-3}$ g/C for Al. The properties of the deposited films, as, for example, its adhesion to the substrate, its crystal structure (the size of microcrystals), etc., may be influenced by the composition of the electrolyte. By this method it is, of course, possible to deposit films only on metallic substrates and the films may be contaminated by substances in the electrolyte.

(*b*) Electroless deposition is based upon a similar principle, but in this case the metal is deposited from the solution by electrochemical processes without the presence of an externally applied field. The rate of deposition depends on the temperature of the bath and in some cases the deposition needs to be stimulated by a catalyst. This method of deposition is used, for example, in the formation of nickel layers on the surface of other metals.

(c) Anode oxidation (anode electrolytic deposition) is used mainly in the formation of films of the oxides of certain metals, including Al, Ta, Nb, Ti and Zr. The oxidized metal is an anode dipped in the electrolyte from which it attracts the oxygen ions. The ions pass through the already formed oxide film by diffusion forced by a strong electric field and combine with the metallic atoms to form molecules of the oxide. Because the growth rate of the film depends exponentially on the intensity of the electric field, the film thus formed is homogeneous owing to the fact that random fluctuations are immediately smoothed out. For anode oxidation it is possible to use either the constant-current or constant-voltage method. Solutions or melts of various salts, or in some cases acids, are used as electrolytes.

Some electrolytes readily dissolve the oxide that is formed so the resulting film is porous and oxidation proceeds through the pores. The thickness of the film is then proportional to the time of oxidation and oxidizing current. An example of this is the oxidation of aluminium in sulphuric or oxalic acid. Other electrolytes (e.g. ammonium citrate or ammonium tartrate) have almost no solvent effect upon the growing film of oxide, so that after a certain thickness is reached − the voltage being kept constant − the oxidation rate falls almost to zero. The final film thickness is then proportional to the applied voltage. This makes it possible to achieve the required thickness without the necessity of additional measurement. It should be pointed out, however, that the film thickness depends very much on the temperature and, to a certain extent, on the kind of electrolyte used.

By this method it is not possible to prepare films of selected thickness. At greater thicknesses breakdown ensues as a result of the excessive growth of the number of charge carriers or the film changes from the amorphous to a microcrystalline state.

The oxide ions needed for oxidation can be obtained not only from an electrolyte, but also from a discharge. Recently anode oxidation in a glow discharge has been tried and it has been found that it is governed by laws similar to those of oxidation in an electrolyte and results in an oxide of a very high quality.

It is necessary in this connection to point out that it is possible to form oxide films by thermal oxidation. On the surface of metals such as aluminium, a thin film of oxide (3−4 nm) forms at room temperature, and the thickness of the film is increased by a rise in temperature. An important practical application is the thermal oxidation of silicon, which gives a film with excellent insulating properties. The oxidation of silicon is often used in semiconductor technology for the passivation of silicon surfaces.

In a process similar to that of oxide formation it is possible at high temperatures to obtain nitrides by reaction with nitrogen (usually in the form of $NH_3$).

(*d*) Chemical vapor deposition is a widely used method in semiconductor technology for the preparation of thin monocrystalline films of high purity. They are often deposited on a substrate of the same material (e.g. Si upon Si), a process called homoepitaxy. When deposited on a different material, the process is called heteroepitaxy. To carry out such deposition, several types of chemical reactions are available. One of them is pyrolysis (i.e. decomposition at high temperature), another is photolysis (i.e. decomposition caused by ultraviolet or infrared light) of gaseous compounds, as for example with the hydrides $GeH_4$ and $SiH_4$ from which pure Ge or Si are obtained, or tetraethylsilicate yielding $SiO_2$ by decomposition, etc. Another method is the reduction of chlorides, e.g. $SiCl_4$ or $SiHCl_3$, either by hydrogen or by a reaction of the type:

$$A + AB_2 \rightleftarrows 2\,AB$$

where $AB$ is a gaseous compound and $A$ the substance it is desired to deposit. The character of the reaction is such that at high temperatures the $AB$ is a stable compound, whereas at lower temperatures the $A$ component separates out. An example of the reaction is

$$2\,SiI_2 \rightleftarrows SiI_4 + Si$$

Chemical transport reactions are used also for preparation of III − V compounds films (i.e. compounds consisting of the elements from the 3rd and 5th groups of the Mendeleyev system, a typical example being GaAs). The principle is as follows. The given substance is treated with, for example HCl or other gas or vapor, to form the volatile compound in which it is transported to a substrate of suitable crystalline material, where it is re-formed, in an appropriate temperature regime, as a thin epitaxial film. Another system used is shown in Fig. 1. In the first zone, reaction of $AsCl_3$ with hydrogen yields vapors of As and HCl. HC, reacts with Ga and forms GaCl and hydrogen; Ga also reacts with As vapors and a thin film of GaAs is produced. Upon the saturation of Ga by As, the final reaction proceeds in the end zone:

$$6\,GaCl + As_4 \rightleftarrows 4\,GaAs + 2\,GaCl_3$$

where GaAs is deposited in crystalline form on the substrate, the other reaction components remaining in gaseous form.

A disadvantage of this system is that As dissolves in Ga; this can be circumvented by a direct reaction of HCl with Ga, followed by the reaction of the resulting chloride with the As vapors transported by the hydrogen flow.

Recently the so-called 'close-spaced' or 'sandwich' method has been developed which utilizes the transport of a substance in gaseous form by means of HCl, water vapor, or iodide vapor, etc., along a very short distance

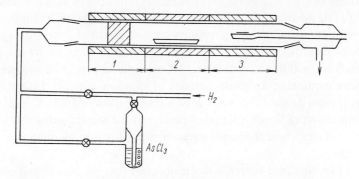

*Fig. 1:* Apparatus for the preparation of GaAs films by transport reaction: (1) As source at a temperature of 425 °C; (2) Ga source, 800 °C; (3) substrate, 750 °C or 900 °C.

(of some 100 μm). The difference in temperature between the source and the substrate of epitaxial layer is maintained at several tens of degrees. Small polycrystals may be used as a source. The efficiency of the method is very high (90%) and it is possible to use it for the formation of single-crystal layers of materials which otherwise had not previously been prepared in the monocrystalline form. It is chiefly employed in the preparation of layers of GaAs, GaP and ternary compounds of the GaAsP type.

(*e*) Liquid phase epitaxial (LPE) growth is a method of depositing semiconducting epitaxial films based on crystallization of semiconducting materials dissolved in a suitable metal. (Metals with a low melting point are usually used, e.g. Sn, In, Pb, Bi, Ga.) A saturated solution is prepared at a high temperature (1000 °C), then gradually cooled. The solution becomes supersaturated and a crystalline phase begins to grow over a given substrate. The solution is then removed by chemical or mechanical means from the substrate. By this process monocrystalline films with a low number of crystal defects may be prepared.

(*f*) Thin films of some high-molecular-weight compounds. The method has been elaborated by Blodgett and Langmuir. A small amount of a high-molecular-weight substance which has polar molecules (e.g. fatty acids or higher alcohols) is dissolved in a volatile solvent and one drop of the solution is sprinkled on the surface of the water. The solvent evaporates and the molecules of the substance diffuse over the surface of the water, all orientated in the same manner due to their polarity. According to their concentration either a 'two-dimensional gas' is formed or a monomolecular film of liquid or solid. Such a film can be lifted up and put upon a plate. Several such films can be piled up on each other and in this way we can gradually form films producing interference colors (i.e. films some hundreds of nanometres thick).

(g) By adsorption it is possible to prepare very thin films (of the order of several monolayers) of, e.g. cesium, on the surface of appropriate photo-cathodes. Another possibility is the preparation of films by means of diffusion from within the substrate. An example is thoriated tungsten, i.e. tungsten containing a certain amount of thorium dioxide, $ThO_2$. During the activation process, i.e. heating to a temperature of 1800 °C, the partial decomposition of the dioxide takes place and the resulting thorium diffuses to the surface, where it forms extremely thin (down to monoatomar thickness) film.

From this brief survey it is already clear that the majority of chemical methods of preparation of thin films are only applicable to a small group of materials. Some of the methods are excellent and are frequently used in semiconductor technology. Nevertheless, the most widely adopted methods of preparation of thin films remain the physical methods, and we shall consider these in the chapters that follow.

# 2.2 Cathode Sputtering

The most important physical methods for the preparation of TF are cathode sputtering and vacuum evaporation. As we shall see, both methods require lowered pressure in the working space and therefore make use of vacuum techniques. We shall deal with practical methods of achieving lower pressures in appropriate places in the text, but detailed information can be obtained from textbooks of vacuum physics and technology [25], [30], [34].

## 2.2.1 Principle of Diode Sputtering

The simplest arrangement for cathode sputtering may be set up as follows (Fig. 2): The material to be sputtered is used as a cathode in a system in which glow discharge is established in an inert gas (e.g. argon or xenon) at a pressure of $10^{-1} - 10^{-2}$ torr and a voltage of several kilovolts. The substrate on which the film is to be deposited is placed on the anode of the system.

It is well known that in the case of glow discharge the potential between the electrodes does not vary evenly and the so-called cathode fall (Fig. 3) results.

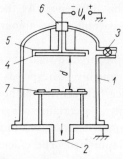

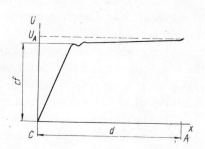

*Fig. 2:* Apparatus for diode cathode sputtering: (1) bell jar; (2) exhaust pipe; (3) inert gas admittance; (4) cathode made from material to be deposited; (5) cathode shielding; (6) high-voltage bushing; (7) substrates.

*Fig. 3:* Variation of potential between cathode $C$ and anode $A$ in a glow discharge: $cf$ — cathode fall; $U_A$ — anode potential.

The positive ions of the gas created by the discharge are accelerated towards the cathode (target) and they arrive there with almost the same speed as they gained in the cathode fall region. The magnitude of the 'normal' cathode fall depends on the sort of gas and on the material of the cathode. In the case of cathode sputtering, we make use of an operating regime in which 'anomalous' cathode fall is produced, which increases with the voltage applied. Under the bombardment of the ions the material is removed from the cathode (mostly in the form of neutral atoms and in part also in the form of ions). The liberated components condense on surrounding areas and consequently on the substrates placed on the anode. We shall return later to the process of removal of cathode particles under ion bombardment.

The amount of the material sputtered, $Q$, in a unit of time is, under constant conditions, inversely proportional to the gas pressure $p$ and anode-cathode distance $d$:

$$Q = \frac{k \cdot V \cdot i}{p \cdot d} \tag{2.1}$$

where $k$ is the constant of proportionality, $V$ is the working voltage, and $i$ is the discharge current. With an increase in pressure or distance there is a corresponding rise in the number of particles which do not reach the substrate because of collisions with other particles. The sputtered amount increases with discharge current $i$ and voltage $V$, which may be explained as follows: There is a certain minimal value of cathode fall under which sputtering does not occur, but above that value the sputtered amount is proportional to the difference between the real fall and critical one.

The symbol $V$ in Eq. (2.1) represents this difference. As the value of cathode fall is directly proportional to the current $i$, the amount $Q$ is thus proportional to $i^2$.

The experimental conditions are usually prepared in such a way that the length of the cathode dark space (which roughly corresponds to the region of cathode fall) is equal to the distance between the cathode and substrate, which for a given pressure occurs only at a certain voltage.

The sputtering rate increases with the atomic mass of ions impinging on the cathode and varies with the materials sputtered. The efficiency of cathode sputtering is given by the coefficient of cathode sputtering, $S$:

$$S = \frac{N_a}{N_i} = 10^5 \frac{\Delta W}{i \cdot t \cdot A} \quad \text{(atoms/ion)} \qquad (2.2)$$

where $N_a$ is the number of atoms sputtered, $N_i$ is the number of impinging ions, $\Delta W$ is the decrease of target mass, $i[A]$ the ionic current, $t[s]$ the duration of bombardment and $A$ the atomic mass of sputtered metal. In

Coefficient of Cathode Sputtering for Some Metals *Table 1*

| Target | Energy of Ar$^+$ ions (keV) | | | | | |
|--------|------|------|------|------|------|------|
| | 0.2 | 0.6 | 1 | 2 | 5 | 10 |
| Ag | 1.6 | 3.4 | | | | 8.8 |
| Cu | 1.1 | 2.3 | 4.2 | 4.3 | 5.5 | 6.6 |
| Fe | 0.5 | 1.3 | 1.4 | 2.0 | 2.5 | |
| Ge | 0.5 | 1.2 | 1.5 | 2.0 | 3.0 | |
| Mo | 0.4 | 0.9 | 1.1 | | | 2.2 |
| Ni | 0.7 | 1.5 | 2.1 | | | |
| Si | 0.2 | 0.5 | 0.6 | 0.9 | 1.4 | |

Table 1, the values of $S$ are given for some metals bombarded by Ar$^+$ ions with various energies (from 0.2 to 10 keV). An example of the dependence of $S$ on the energy of the ions is shown in Fig. 4.

The minimal (threshold) energy at which the sputtering still occurs varies for different metals and ions from 40 to 130 eV. The threshold energy decreases, i.e. the sputtering increases, — provided other conditions are the same — as the atomic number increases in each group of the Mendeleyev system (Fig. 5).

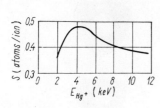

*Fig. 4:* Sputtering yield as a function of the impingement energy of Hg$^+$ ions.

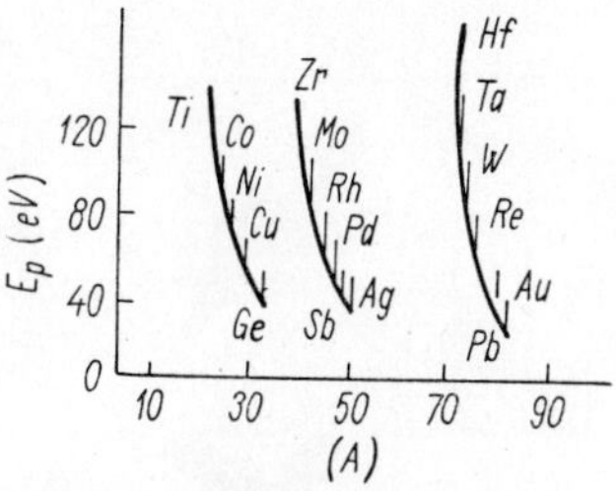

*Fig. 5:* The dependence of threshold sputtering energy $E_p$ of Hg$^+$ ions on the atomic number $A$ of target material.

If the sputtered material is in polycrystalline form, the angular distribution of sputtered particles leaving the cathode is

$$i_a = i_0 \cdot \cos(\alpha) \tag{2.3}$$

where $i_a$ is the current of particles ejected at angle $\alpha$, $i_0$ is the current of particles ejected perpendicular to the surface. When a single crystal is sputtered, the particles are mainly emitted along certain definite directions (Fig. 6).

X-ray analysis has established that the directions of most intensive sputtering correspond to the directions of the most closely packed arrangement of atoms in a crystal lattice.

Various hypotheses have been advanced to account for the mechanism of cathode sputtering. According to Hippel, the local heating caused by the impact of ions on a particular site effects a thermal evaporation of particles, even when the mean temperature of the cathode remains low. Nowadays, however, the most probable explanation is considered that advanced by Stark, developed by Langmuir and put into mathematical form by Henschk *et al.* According to this hypothesis, the governing mechanism is that of momentum transfer from impinging particles to those of the crystal lattice. In the simplest case it may happen that the momentum is transferred from the ion directly to the emitted particle. This type of transfer is, however, relatively infrequent and contributes to the total sputtered amount by only a few percent. A large number of ions, however, penetrate rather deep into the cathode and their momenta are transferred successively from one atom to another (in a manner analogous to colliding billiard balls). It has been found that in certain crystallographic directions so-called 'focussed collision sequences' arise which can transfer the momentum over a long chain of atoms. The hypothesis is substantiated by experiments with the sputtering of single crystals, which show us in which crystallographic directions the momentum transfer is most efficient (see Fig. 6). Other data appear to

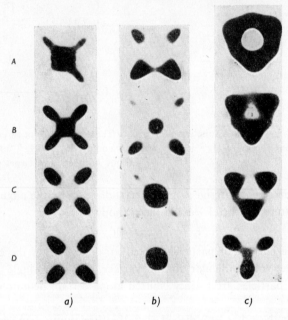

*Fig. 6:* Patterns prepared by sputtering of Cu single crystal: (a) Cu (100) planes, by 500 eV ions of: A—Ne$^+$, B—Ar$^+$, C—Kr$^+$, D—Xe$^+$; (b) Cu (100), Kr$^+$ ions of energy: A — 500 eV, B — 1000 eV, C — 1500 eV, D — 2000 eV; (c) Cu (111) plane, Kr$^+$ ions of energy: A — 250 eV, B — 500 eV, C — 100 eV, D —1500 eV.

contradict the hypothesis of local heating too. For example, if such heating really occurs, then thermo-electrons would also have to be emitted from the heated point. However, such emission has not been observed. Nor does the threshold energy of cathode sputtering correspond to the latent heat of evaporation of the substance, always being higher. Finally, the energy distribution of emitted particles is also very different from that of evaporated ones (Fig. 7), the mean energy being roughly ten times higher. All these experimental facts indicate that the mechanism of cathode sputtering is connected with transfer of momentum and not with local heating.

The theory affords rather complicated expressions for the sputtering coefficient that contain mostly empirical parameters and are based on many simplified assumptions.

For example, Keywell considers only simple elastic collisions characterized by accommodation coefficients. He introduces a threshold energy of the displacement of an atom in a lattice, the probability that the displaced atom reaches the surface and the probability that the ion (provided $m_{ion} <$ $< m_{evaporant}$) will be rebounded into the vacuum by the first collision. By a suitable choice of the parameters, good agreement with experiments may be attained, especially for ion energies $< 5$ keV.

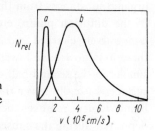

*Fig. 7:* Velocity distribution of Cu particles: (a) in evaporation; (b) in cathode sputtering ($N_{rel}$ — relative number of particles with corresponding velocity).

For higher energies $(5-50$ keV$)$, the theory advanced by Kistemaker *et al.* is applicable. It is assumed that only the first collision of ions with a target atom, taking place near the surface, is important. The probability of the collision is inversely proportional to the mean free path $\lambda(E)$ of ions in the material. Thus, the sputtering coefficient is also inversely proportional to the path, besides being directly proportional to the mean energy transferred to the atom by the first collision which is approximately

$$\sim E \frac{m_i - m_a}{(m_i + m_a)^2}$$

The most general theory is that of Sigmund (*Phys. Rev.*, **184** (1969), 383), which is applicable for both amorphous and polycrystalline substances and is based on the solution of the Boltzmann transport equation.

### 2.2.2 Some Special Systems of Cathode Sputtering

The technique of cathode sputtering which we have just described, i.e. the normal diode system, has certain shortcomings.

As a consequence of the relatively high pressure in the system the surface of the substrate is also bombarded by molecules of the gas (see Chapter 3) in addition to the sputtered particles. The gas molecules therefore are also embedded in the growing film. Hence, it is not possible to obtain films of the extreme purity necessary for many purposes. This is one of the

24

main reasons why cathode sputtering was replaced for a time by evaporation; the technique was considered to be too 'dirty'. Since about 1960, however, cathode sputtering has been making a comeback as several modifications were developed which made it possible to grow films of high purity and defined structure. One of the methods of ridding the surface of deposited films of impurities is that of an asymmetric alternating voltage. In the first half-cycle the substrate is charged positively and is bombarded by sputtered particles and molecules of the working gas and of various impurities (e.g. organic substances from the materials of vacuum seals, from working fluids of the diffusion pump, etc.). During the second half-cycle, the substrate is negatively charged and is bombarded by ions from the discharge cloud; these ions now have lower energy because the amplitude of the voltage is smaller in the second half-cycle. As the binding energy of impurities on the surface is usually lower than the binding energy of sputtered material, the bombardment removes these impurities and thus the sputtered particles are deposited during the subsequent half-cycle on a cleaned surface.

Similar and even more successful results can be achieved when the samples on the conducting substrate are insulated from the anode and a negative bias voltage (about 100 V) is applied relatively to the anode because in such a set-up there occurs a continuous cleaning of the surgace by the ions from the discharge space and so it is possible in this way to obtain very pure films.

The purity of a growing film is endangered by admixtures in the inert gas used, namely, by admixtures of active gases. To reduce the concentration of the active gases a gettering is used. The gettering is a process during which

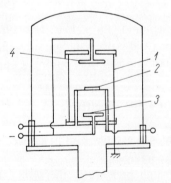

*Fig. 8:* Apparatus for getter cathode sputtering: (1) protective cylinder; (2) substrate; (3) auxiliary cathode; (4) cathode.

the active gases are chemically bounded with atoms of a metal — 'the getter' — which is evaporated or sputtered in the given space and forms a coating on the walls (barium is used for an improvement of the final vacuum in tubes). In the getter sputtering, the getter is the sputtered material itself. The bell jar is divided into two parts, the gettering and the working region, in such a way that the gas which enters the working region has to pass through the gettering region and is thus to a considerable degree deprived of active admixtures (Fig. 8).

On the other hand, it is the reaction with active gases which is employed in reactive sputtering. During the process of cathode sputtering a chemical reaction occurs of cathode material with active gas (e.g. oxygen, nitrogen, $H_2S$), which is intentionally added to the working gas or which may be the working gas itself.

At higher pressures the mechanism probably consists of reaction on the cathode leading to the formation of the relevant compound (e.g. oxide, nitride, sulphide), which is then sputtered by ions. At lower pressures the reaction proceeds perhaps only after the gasification.

Reactive sputtering is suitable for the preparation of thin films of a great number of compounds, namely, of those which dissociate by heating and which cannot therefore be evaporated or of those having a high melting point.

In the cathode sputtering of insulators, a difficulty arises because of a charging of the surfaces. The difficulty is removed by high-frequency cathode sputtering. The dielectric which is to be sputtered is placed with a conducting substrate in the plasma of a low-pressure glow discharge and then a high-frequency voltage is applied to it, the frequency being of the order of MHz. The charge formed by ions on the dielectric surface in one half-period is compensated by electrons from the plasma in the subsequent half-period.

In this way it is possible to attain current densities of 20 mA/cm² and to sputter materials such as silica, diamond, etc.

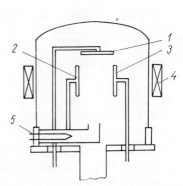

Fig. 9: Apparatus for triode cathode sputtering:
(1) auxiliary anode; (2) cathode; (3) anode; (4) magnetic coil; (5) auxiliary heated cathode.

### 2.2.3 Low-Pressure Methods of Cathode Sputtering

As we have already noted, the sputtering rate is directly proportional to the discharge current. High discharge currents can be achieved only at sufficiently high pressures. On the other hand the use of higher pressures causes the shortening of the mean free path of particles, and this leads to the reduction in the number of particles impinging on the substrate because of more frequent collisons with the gas molecules. The higher pressure also increases the danger of occlusion of the gas in the film. Efforts to lower the working pressure, while keeping the same current density, have resulted in systems employing various means to enhance the ionization effect with simultaneous decrease in the working pressure.

One method is the introduction of an auxiliary thermocathode which emits electrons independently of pressure and represents the requisite agent of ionization (Fig. 9). The method is known as triode cathode sputtering. The ionization is enhanced still further by application of a magnetic field,

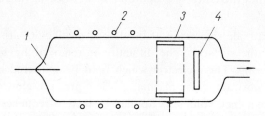

*Fig. 10:* Schematic representation of sputtering with auxiliary high-frequency glow discharge: (1) anode; (2) high-frequency coil; (3) ring cathode; (4) substrate.

which bends the paths of electrons so that they remain a longer time in the discharge and create more ion pairs. A convenient arrangement is one in which the gas is admitted to the region of auxiliary discharge, which is separated from the main recipient by a shutter, so that there is a considerable difference in the pressure. The working pressure in the recipient is $10^{-4}$ to $10^{-3}$ torr, while the pressure in the region of the auxiliary discharge is about twenty times higher.

At lower pressures a sufficient level of ionization can also be achieved by the use of an auxiliary high-frequency discharge, a schematic picture of such a system being shown in Fig. 10. High-frequency sputtering employs a high-frequency voltage between the sputtered target and the substrate, whilst a d.c. voltage is applied between a cathode and an anode and a magnetic field of proper structure improves the uniformity of film thickness (see Section 2.2.2). This system is similar to that shown in Fig. 9

with the difference that a high-frequency voltage is applied between electrode 2 and 3.

In conclusion something should be said about the advantages of cathode sputtering. In comparison with evaporation *in vacuo* there is no danger of film contamination by the evaporation source (the container or holder of the material to be evaporated). It is possible to sputter substances which have very high melting points and are only evaporated with difficulty, and even various alloys and compounds without impairment of the stoichiometric ratio of their components.

The deposition rate can be well controlled by current density and voltage. If the apparatus is to provide films whose purity is to compare with that of films evaporated in ultra-high vacuum (see Section 2.3), the preliminary technological process of evacuation and degassing before admitting the appropriate working gas should be the same as that usual in ultra-high vacuum apparatus. The requirements for the degassing of materials are even more strict in this case because some parts of the apparatus may be heated during sputtering.

We have not so far mentioned the actual process of formation (condensation) of films. The process has many features in common with condensation of evaporated films. There are, of course, some substantial differences, to which we shall separately call attention in the chapter on the mechanism of film formation. One of the main causes of these differences is the difference in the energy of particles, which are in the region of tenths of eV for the evaporated ones while the mean energies of the sputtered ones are of the order of several eV with a marked occurrence of particles with energies of several tens of eV. A practical consequence of this is that as a rule adhesion to a substrate is better for sputtered films than for those evaporated *in vacuo*.

By suitable arrangement of the discharge space and particularly by the utilization of a magnetic field of appropriate configuration, it is possible to achieve the formation of films of large area and of very uniform thickness.

The vacuum systems used for cathode sputtering are analogous to systems used for evaporation (see Section 2.3.21). They are complemented by a gas reservoir and a valve (usually a needle valve) for controlled admittance of the gas into the working chamber. A steady working pressure is obtained by regulation of the admittance rate at a given pumping speed. This speed is sometimes decreased during the working cycles proper to reduce the consumption of inert gas.

## 2.3 Vacuum Evaporation

### 2.3.1 Physical Foundations

Vacuum evaporation is currently the most widely used method for the preparation of thin films. The method is comparatively simple but it can in proper experimental conditions provide films of extreme purity and, to a certain extent, of pre-selected structure.

The process of film formation by evaporation consists of several physical stages:

(1) transformation of the material to be deposited by evaporation or sublimation into the gaseous state;

(2) transfer of atoms (molecules) from the evaporation source to the substrate;

(3) deposition of these particles on the substrate;

(4) their rearrangement or modifications of their binding on the surface of the substrate.

It is a known fact that atoms or molecules are liberated by heating from every material in its solid or liquid phase and that in a closed system a certain equilibrium pressure, which is called the saturated-vapor pressure is established at a given temperature. The dependence of the pressure on temperature is rather strong as may be seen in Fig. 11a, where the curves for several important materials are given.

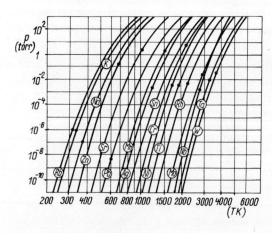

*Fig. 11a:* Temperature dependence of vapor pressure for various materials.
● – melting point

The liquid → vapor transformation is called evaporation, the solid → vapor one is called sublimation. Both processes are used in thin-film deposition.

In order that a given molecule may leave the surface of a material, it is necessary that the kinetic energy corresponding to the component of velocity perpendicular to the surface be higher than the energy $L_v$ needed to surmount attracting intermolecular forces. The kinetic energy is determined by the thermal motion of the molecules and thus the number of particles fulfilling the inequality increases with temperature. The evaporation therefore occurs at the expense of the internal energy of the body. To prevent a decrease in its temperature, heat has to be supplied (internal evaporation heat). Moreover, there is additional work done during evaporation in consequence of the expansion of volume occurring during the transition into gaseous form (external evaporation heat).

In a state of equilibrium (i.e. when the vapor pressure equals the saturated-vapor pressure) the basic quantities are interrelated by the Clapeyron-Clausius equation

$$\frac{\mathrm{d}p}{\mathrm{d}T} = \frac{L_v}{T \cdot \Delta v} \tag{2.4}$$

where $p$ is the pressure, $\Delta v$ the volume change and $T$ the absolute temperature.

The number of particles with molecular weight $M$ evaporated at a temperature $T$ (in equilibrium) per unit time from a square centimeter can be determined using elementary kinetic theory and is given as

$$N_e = \frac{p_e}{\sqrt{(2\pi MKT)}} = 3.513 \cdot 10^{22} \frac{p_e^{torr}}{\sqrt{(MT)}} \tag{2.5}$$

where $p_e$ is the vapor pressure.

If the system is not in equilibrium and there is a relatively lower temperature in some part of it, the vapor will condense in this part and conditions will thus be established for a transfer of material from evaporation source to a colder substrate. The deposition of a film by evaporation is thus essentially a nonequilibrium process.

The liberated particles travel in space with their thermal velocities along a straight line until collision with another particle. To ensure a straight path for them between the source and substrate, particle concentration in the space must be low, i.e. the space must be sufficiently exhausted. The portion of particles scattered by collisions with atoms of residual gas is proportional to $1 - \exp(-d/\lambda)$ where $d$ is the source-substrate distance and $\lambda$ is the mean free path of the particles.

For air at 25 °C $\lambda$ is in the range of 50 to 5000 cm at pressures ranging from $10^{-4}$ to $10^{-6}$ torr; it is therefore obvious that at the normal apparatus dimensions (of the order of tens of centimeters), it is necessary to use pressures of $10^{-5}$ torr if a considerable dispersion of evaporated particles is to be prevented.

However, the residual gases in evaporation systems have still another effect which in some cases necessitates the use of lower working pressures. The substrate is bombarded not only by particles of the evaporated substance, but also by particles of residual gases.

The number of particles evaporated in unit time from a unit area of surface of a material with molecular weight $M$ is given by equation (2.5).

The number of the particles which actually impinge on the substrate also depends on the geometric configuration of the system, but is nevertheless proportional to the quantity $N_e$. Simultaneously with those particles, a certain number of residual gas particles arrive at the substrate, from which a fraction, determined by the so-called sticking coefficient $s$, can be adsorbed on the substrate. The number of these impinging particles is

$$N_a = 3.513 \cdot 10^{22} \frac{p_a}{\sqrt{(M_a \cdot T_a)}} \qquad (2.6)$$

Various Kinetic Data for Air                                          Table 2

| Pressure (torr) | Mean free path (cm) | Collision rate $(s^{-1})$ | Impingement rate $(s^{-1} cm^{-2})$ | Number of monolayers $(s^{-1})$ |
|---|---|---|---|---|
| $10^{-2}$ | 0.5 | $9 \times 10^4$ | $3.8 \times 10^{18}$ | 4400 |
| $10^{-4}$ | 51 | 900 | $3.8 \times 10^{16}$ | 44 |
| $10^{-5}$ | 510 | 90 | $3.8 \times 10^{15}$ | 4.4 |
| $10^{-7}$ | $5.1 \times 10^4$ | 0.9 | $3.8 \times 10^{13}$ | $4.4 \times 10^{-2}$ |
| $10^{-9}$ | $5.1 \times 10^6$ | $9 \times 10^{-3}$ | $3.8 \times 10^{11}$ | $4.4 \times 10^{-4}$ |

where $p_a$ is the equilibrium gas pressure at the temperature $T_a$. Table 2 displays some relevant quantities for air at the temperature of 25 °C and Table 3 gives the values of the coefficient $K$, defined as the ratio of the number of residual gas molecules impinging on 1 cm² of surface per second to the number of evaporation-deposited particles, at various evaporation rates assuming the coefficient $s$ equals unity. It can be seen from Table 3 that under the usual conditions the ratio is rather unfavorable even when taking into account that the coefficient $s$ is in reality smaller than unity for residual gas molecules while often approaching unity for evaporated materials. From this it follows that the requirement of extreme purity of the film necessitates the use of pressures in the region of ultra-high vacuum, i.e. $<10^{-8}$ torr.

Values of the Constant $K$       *Table 3*
for Various Pressures $p$ and Evaporation Rates $R$

| $p$ (torr) | \multicolumn{4}{c}{$R$ (nm/s)} |
| --- | --- | --- | --- | --- |
| | 0.1 | 1.0 | 10.0 | 100.0 |
| $10^{-9}$ | $10^{-3}$ | $10^{-4}$ | $10^{-5}$ | $10^{-6}$ |
| $10^{-7}$ | $10^{-1}$ | $10^{-2}$ | $10^{-3}$ | $10^{-4}$ |
| $10^{-5}$ | $10$ | $1$ | $10^{-1}$ | $10^{-2}$ |
| $10^{-3}$ | $10^{3}$ | $10^{2}$ | $10$ | $1$ |

The purity and the morphology of films can be influenced by residual gas pressure, evaporation rate, and also by the temperature and the structure of the substrate (the coefficient $s$ depends on the two latter quantities). This means that for the formation of films with reproducible properties, it is

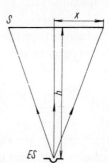

*Fig. 11b:* Schematic illustration for the determination of thickness distribution: ES — evaporation source; S — substrate.

necessary that these parameters be constant and, of course, measurable. As will be shown later, the morphology of a film, especially the degree of contamination, substantially affects the electric, magnetic and other properties of films.

It has been already said that the number of vacuum-deposited particles depends on the geometrical configuration of the system, i.e. on the shape and the relative situation of the source of evaporation and the substrate. If the source is approximately point-like and the substrate is plane (Fig. 11b), the film thickness (deposited in a certain time) decreases with the square of the distance, so the greatest thickness $t_0$ is obtained in the centre, i.e. at the distance $h$, and diminishes with $x$, the distance from the centre, according to the formula

$$\frac{t}{t_0} = \frac{1}{[1 + (x/h)^2]^{3/2}} \tag{2.7}$$

If the substance is evaporated from a comparatively small emission plate onto a parallel plane, the relevant formula is

$$\frac{t}{t_0} = \frac{1}{[1 + (x/h)^2]^2} \tag{2.8}$$

Two further stages of film formation, namely condensation on the substrate and particle rearrangement, are discussed in Sections. 4.2 and 4.3.

In conclusion we shall deal briefly with the practical problem of evaporation of multicomponent materials such as alloys and compounds. Here there is the danger that the evaporated substance will dissociate at high temperature and that the components, having different saturated-vapor pressures, will evaporate at different rates, so that the composition of the film will differ from that of the source material.

Different reactivities of the individual components with the evaporation source can add further problems.

If the substance consists of A and B components with molecular weights $M_A$ and $M_B$ and equilibrium pressures of (saturated) vapors $p_A$ and $p_B$, it follows from relations (2.5) and (2.6) that the ratio of the numbers of deposited particles of the components is

$$\frac{N_a}{N_B} = \frac{C_A}{C_B} \cdot \frac{p_A}{p_B} \sqrt{\frac{M_B}{M_A}} \tag{2.9}$$

where $C_A$ and $C_B$ are the fractions of the components.

Their ratio in the vapor is thus generally different from the stoichiometric ratio in the parent material. A correction should be made in the ratio for the condensed state to allow for the fact that the individual components and the compound to be deposited have in general different condensation coefficients depending on their adsorption energies on a given substrate. The coefficients influence the probability of mutual collisions of the individual particles on the substrate and hence the formation of the desired molecules. Best results are achieved when the adsorption energy of the desired compound AB is higher than those of both components A and B. The substrate temperature can then be set so that the compound condenses on it but not the components. If such conditions are not present, the given compound cannot be obtained in stoichiometric composition by evaporation from a single evaporation source. The higher the temperature, the smaller the difference between the compositions of the evaporated film and the parent material. As the individual components may react with the substrate material, chemically inactive substrates are recommended for use in such cases.

For some multicomponent compounds (e.g. compounds from the second and sixth, or third and fifth, groups of the Mendeleyev system) evaporation is conducted from separate evaporation sources (each having a different temperature) and the substrate itself is kept at an elevated temperature. This so-called three-temperature method gives good results, but the temperatures are of critical importance.

We should mention here the getter evaporation in which (as in the getter sputtering) an active component of residual gas reacts chemically with the deposited material so that its content in the residual atmosphere is reduced. At first, the component is evaporated in a half-closed space so that the evaporated material cannot condense on substrates but on the walls of the evaporation chamber, which are usually cooled. Only after the partial pressure of the reactive components of the residual gas drops sufficiently is the shutter shielding the substrate opened and the evaporation begin to take place. In this way, films equivalent to those prepared in ultra-high vacuums may be deposited in a normal commercial apparatus.

The reaction of deposited material with an active component of the gas in the evaporation chamber is on the other hand utilized in so-called reactive evaporation. The reaction proceeds mostly on the substrate in the form of chemisorption since collisions in the evaporation space are very improbable as a result of the pressures used, whereas the numbers of atoms of the deposited material and of gas atoms impinging on a unit surface per unit time are comparable (for example, during the deposition of aluminium at a rate of 1 nm/s in oxygen at a pressure of $10^{-5}$ torr). In this manner it is thus possible to form films of some compounds, e.g. oxides of the $Al_2O_3$ type, which are otherwise evaporated with difficulty.

### 2.3.2 Experimental Techniques

#### 2.3.21 Evaporation Apparatus

The systems for the evaporation *in vacuo* have to meet some special requirements:

(1) Sufficiently low threshold pressure.

(2) Sufficiently fast achievement of threshold pressure from atmospheric.

(3) Working chamber uncontaminated by organic vapors.

(4) Spacious and easily accessible working chamber.

(5) The possibility of installation of a sufficient number of electrical feedthroughs into the working chamber and the possibility of transmission of motion into the vacuum.

The first two requirements necessitate the use of high-vacuum pumps with sufficient pumping speeds and the shortest and widest possible connecting piping between the pump and the exhausted space. The most widely used are metal-oil-diffusion pumps with a high-quality mineral or silicone oil as a working fluid, pre-exhausted by rotary pumps. The pumping speed is usually set at a level of several hundreds of liters per second. The penetration of oil vapors into the working chamber is prevented by insertion of suitable traps (cooled by water or liquid nitrogen) above the pump. By placing a valve between the pump and the exhausted chamber the working cycle can be shortened, especially when there is a direct pre-exhaustion of the recipient (or by-pass), but only at the price of increased vacuum resistance and thus lowered pumping speed.

The achievement of low threshold pressures requires, of course, sufficient tightness of the system and the use of low-vapor-pressure materials which can be degassed. The best construction material is polished stainless steel. In ordinary apparatus rubber or silicon rubber gaskets are used; in a more demanding apparatus the seals are made from Viton, which can be degassed at temperatures up to 150 °C; in ultra-high vacuum apparatus only metal seals should be used, mostly copper or gold ones.

A sufficiently spacious and easily accessible working chamber usually consists of a bell jar (usually of metal, in the older apparatus a glass one) mounted upon a metal baseplate, equipped with a hand- or hydraulically-operated lift. The lower part has the form of a collar with many feedthroughs (electrical and mechanical — for the transmission of rotary or translational motion into the vacuum) and with branches for connections to vacuum gauge, mass spectrometer or other auxiliary devices. The bell jar is usually furnished with glass windows for visual inspection. In addition there is often a needle valve for controlling the gas inlet and sometimes also an optical monitor of the deposited film thickness. Some types of ultra-high vacuum deposition systems have a double wall bell jar, the inner jacket of which is made from a very thin sheet and can be degassed by direct passage of a current. The space between the jackets is exhausted by an auxiliary pump. The bell jar usually has a copper pipe welded to the outer jacket through which hot or cold water can be driven according to need. The layout of such an apparatus is shown in Fig. 12a.

Modern industrial systems usually have an immovable cylindrical or angular bell jar with one tilting sidewall. Film substrates and evaporation elements are placed inside through this port, usually on a truck equipped with a frame which may, if needed, provide motion (rotation) in the vacuum.

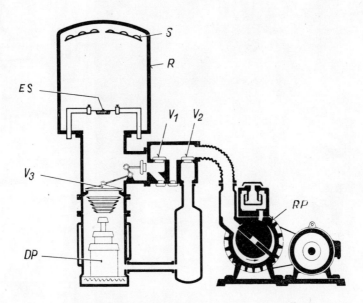

*Fig. 12a:* Layout of vacuum evaporation system: RP — rotary pump; $V_1$, $V_2$, $V_3$ — valves; DP — diffustion pump; R — bell jar; ES — evaporation source; S — substrates.

For deposition in ultra-high vacuum other pumping systems of exhaustion are used to ensure that organic vapors will not be present. As auxiliary (prevacuum) pumps, sorption pumps with molecular sieves are employed. After pre-degassing, the sorption pumps, operating at liquid nitrogen temperature, reduce the pressure to the range of $10^{-4} - 10^{-5}$ torr by adsorption of a large quantity of gas on their inner surfaces. Further reduction of pressure is achieved by means of an ion-getter pump (so-called Hall pump). The principle of the pump is similar to that of a Penning discharge vacuum gauge. The titanium is sputtered by a glow discharge established in a magnetic field between two titanium cathodes and a grid anode. The sputtered titanium combines with gas molecules by chemisorption and, in addition, the physically sorbed particles are buried under subsequent sputtered layers. By such pumps, it is possible in a perfectly built and degassed system to achieve pressures in the region $10^{-10} - 10^{-11}$ torr. During

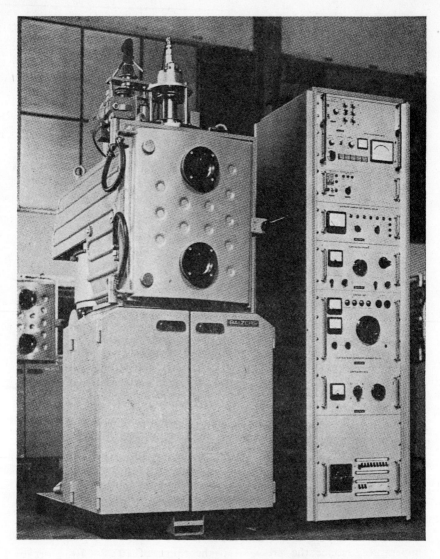

*Fig. 12b:* A high-vacuum evaporation system for industrial production of thin films for optics and electronics (by permission of Balzers AG).

the evaporation proper an increase in pressure always takes place owing to the liberation of gas from the evaporated substance even if it is pre-degassed (as degassing is never complete). An additional pump is therefore used, namely, a titanium sublimation pump, in which titanium is sublimated and through sorption binds the gas molecules. The pumping speed depends on the evaporation rate of titanium and can easily reach values in the range of thousands liters per second. The pumping speed is increased by cooling the condensation surface to a low temperature. Such a system consitutes a half-way stage to cryogenic pumps, which utilize the condensation of gases on a surface cooled to a very low temperature by e.g. liquid helium.

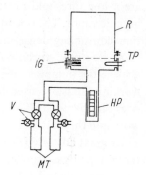

*Fig. 13:* Schematic diagram of ultra-high-vacuum apparatus: MT — molecular traps; HP — Hall pump (ion-getter pump); TP — titanium sublimation pump; IG — ionization gauge; R — bell jar.

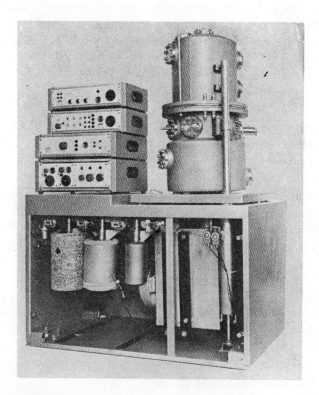

*Fig. 14:* Ultra-high-vacuum evaporation system Balzers BA 330 U with sorption, ion and titanium evaporation pumps (reprinted with permission from Balzers).

The bell jar and high-vacuum valve of apparatus of this type are constructed in such a way that it is possible to degas them by bakeout in an oven (at temperatures $\sim 400$ °C). The seals of the high-vacuum chamber are generally made from metal. The scheme of such apparatus is shown in Fig. 13, and an example is given in Fig. 14. The advantages of this apparatus are a low threshold pressure and a residual gas which contains almost no organic vapor. The operating cycle is comparatively long due to the necessary time-consuming degassing.

Recently, turbomolecular pumps have also come into use. These achieve the pumping effect by rapid rotation of blade wheels. No sealing or working fluids are used in them and they thus provide a 'clean' vacuum; degassing is possible so that they may be employed for ultra-high vacuum work and they have a constant pumping speed over a wide range of pressures. Their great advantage over ion-sorption pumps is that they actually remove the pumped gas from the apparatus. In the titanium ion-getter pumps the gas is only bound inside the system and may therefore be liberated later. Moreover, with ion-getter pumps a number of chemical reactions occur which may lead to the formation of lower hydrocarbons. All these inconveniences are absent in turbomolecular pumps, which, however, are not so easy to operate.

### 2.3.22 Substrates and Their Preparation

The demands made on the substrates of thin films follow from their purpose: the substrate serves as a mechanical support for the film and in electronic applications it usually serves also as an insulator. The need for long-term stability in thin film substrates makes it imperative that no chemical reactions occur which could change the properties of the film. The substrate must therefore fulfill certain requirements as to mechanical strength and there must be adequate adhesion of the film to the substrate, not only at normal temperature but also during relatively large temperature changes. These changes may arise during the preparatory stage of operation (degassing at high temperature), during the deposition of the film (onto heated or cooled substrate) and sometimes during the operation of the thin film system. A high dielectric strength is also required. To ensure constant temperature of the surface and sufficient heat removal during the operation of electronic elements, an appropriate heat conductivity is necessary. Further, to form the films with defined and reproducible electrical and other parameters, the surface of the substrate should be flat and smooth.

These characteristics should be accompanied by a number of practical requirements, such as the possibility of vacuum processing and the availability and price of the material. In some applications, even its weight may be an important factor. Absence of contamination is, however, necessary in all cases.

It is, in general, possible to say that there is no material that would satisfy all these requirements. The most widely used substrates for poly-crystalline films are glass, fused silica and ceramics (mostly based upon $Al_2O_3$). Organic materials that have also been tested (e.g. Mylar or Teflon), have the advantage of low specific weight, but they cannot endure a high temperature and are used therefore only in special cases. For a single-crystal epitaxial growth the most frequently used materials are the single crystals of alkali halides, silicon, germanium, sapphire and mica.

The alkali content of different glasses is very important, especially that of sodium. Often used is Pyrex (composition: 80.5% $SiO_2$, 12.9% $B_2O_3$, 3.8% $Na_2O$, 2.2% $Al_2O_3$, and 0.4% $K_2O$) or fused silica, which besides its chemical inactivity is temperature-resistant to a considerable degree. The substrates with higher alkali content are often the cause of instability in the electrical and other properties of the films. For example, $Na_2O$ is practically immobile up to 4% concentration; at higher concentrations, however, and especially at elevated temperature and intensity of electric field, it can travel rather easily in the glass. Together with moisture, it forms a layer of high conductivity on the surface and causes electrolytic corrosion of the film especially in the neighborhood of negative contacts. From this point of view, a glazed ceramic is a better material even when the alkali content of the glazing is about the same. The reason lies in the better heat conductivity of the material, which means a lower operating temperature of the surface with the same electric current.

The surface of glass can be made relatively flat and smooth. The roughness is measured by an apparatus called a Talysurf, which consists essentially of a very thin and hard tip which moves over the surface being investigated, its vertical movements being amplified and transmitted to a recorder. The device can resolve irregularities 5 nm high and 2.5 µm long. The irregularities in glass can be reduced to a 5-nm height by optical polishing or fire polishing (i.e. heating the surface to a softening tempera-ture). The smoothest surfaces can be obtained by using polished fused quartz and polished single crystals of sapphire. For comparison, the profiles of some surfaces are shown in Fig. 15. It should be noted, however, that polishing is an expensive operation, which can moreover produce undesirable stresses. Some glasses, however, have sufficiently smooth surfaces even after drawing (with accidental irregularities in the range of 100 nm). For further smothen-ing and chemical passivation, an SiO layer is sometimes deposited on the substrate.

40

Ceramic substrates, which have an advantage over glass ones in their higher mechanical and thermal resistance and thermal conductivity, are of lower quality in regard to the smoothness of the surface, which is affected by the sintered-grain structure. But again, a considerable improvement is achieved by polishing (see Fig. 15).

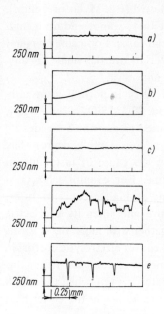

*Fig. 15:* Surface profiles of some substrates: (a) soft glass drawn sheet; (b) another kind of drawn glass; (c) optically ground glass; (d) ceramic (96% $Al_2O_3$) after annealing; (e) same ceramic polished.

The cleaved surfaces of single crystals (in air or vacuum) may be absolutely smooth (i.e. on an atomic scale) as, for example, in the case of mica. The surface of alkali halides shows steps which may be made visible by the so-called decoration technique, i.e. deposition of ultrathin metal film. The steps act as nucleating centres (see Section 4.2) on which, for the most part, the metal condenses.

An NaCl single-crystal surface decorated by a thin film of Au is shown in Fig. 16. The cutting of a crystal in vacuum (often required in order to keep the surface uncontaminated by air) in some cases leaves a fine dust on the surface which cannot be removed easily and which may also serve as undesirable nucleation centres.

For many applications, an important factor is the surface conductivity. Its measurement is carried out by the arrangement illustrated in Fig. 17. This records the conductivity between two deposited contacts with a separation of 10 mm. The potential fall across a resistance, $R$, due to the current is measured by an electrometer, $E$, and displayed on a graph plotter, $GP$, together with the temperature measured by a thermoelement, $T$. In fact, we do not measure the surface conductivity, but a combination of the surface and bulk conductivities, which is important from the practical point of view.

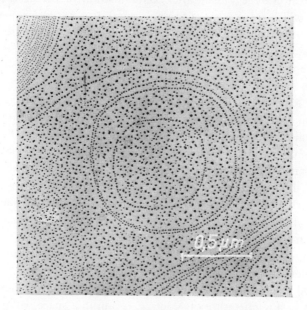

*Fig. 16:* Surface of a single crystal of NaCl with sputtered Au film (reproduced with the permission of Dr Šimečková).

The resistances thus measured (in dry air) are in the range of $10^{14}$ to $10^{16}$ Ω for the substrates in common use.

As we have already noted it is necessary for the constancy of the film properties and for its proper adhesion that the surface of the substrate be

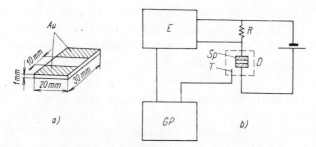

*Fig. 17:* Measurement of surface conductivity: (a) schematic pattern of the specimen; (b) schematic measuring circuit; $E$ — electrometer, $R$ — resistance standard, $O$ — heat bath, $Sp$ — specimen, $T$ — thermocouple, $GP$ — graph plotter.

absolutely clean and uncontaminated. This condition is fulfilled for surfaces of single crystals obtained in vacuum, but cannot be met in other cases. The surface of glass is relatively clean after drawing, but there is no way of attaining that cleanness again after contamination of the glass (except by remelting, which is the basis of fire polishing). It is not always necessary to meet these extreme requirements of surface cleanness because we often deal in practice with the removal of only certain detrimental contaminants. But no agents that could cause mechanical or chemical damage should be used, i.e. no abrasive substances or strong chemicals.

The recommended procedure for cleaning is as follows: application of ultrasound in a solution of ionic detergent, thorough rinsing in deionized distilled water, degreasing in a vapor of pure alcohol and finally drying in a stream of dry and filtered nitrogen. A very efficient and often used method for decontamination is ion bombardment (by a beam of ions or in a glow discharge). It is necessary to use low ion energies (up to about 0.5 kV) and low current densities ($< 100\ \mu A/cm^2$) in order to avoid surface damage. The bombardment should be preceded by degassing and annealing. When using this method of cleaning, it is important that the space be devoid of organic vapors which would form a polymer film over the sample during bombardment.

Good results can be achieved with glass substrates by heating them in vacuum to a temperature of $\sim 300\ °C$ (again in a space without organic vapors) and combining this procedure with the above-mentioned 'wet' treatment.

The substrate often has to be heated or cooled during a deposition of film. For the heating, tungsten or tantalum heat sources are placed under the substrate, or a heater in the form of nichrome wire inserted between two mica sheets is employed. Heating by electron bombardment is also used. For cooling, cooling mixtures are utilized, or, if need be, liquid gases — using a copper block. The substrate is cemented to the block with graphite or silver paint to obtain the best possible thermal contact. For temperature measurement, TF thermocouples deposited directly on the substrate are recommended because a low thermal conductivity of the substrates could lead to considerable temperature differences in the substrate.

### 2.3.23 The Most Important Materials for Evaporation

A great diversity in applications of thin films necessitates the use of various evaporants: metals (pure and alloys), semiconductors and dielectrics, chemical elements and compounds. The means have to be provided to heat the material to the temperature adequate for the required rate of evaporation.

The temperature corresponding to the saturated vapor pressures of $10^{-4}$ to $10^{-2}$ torr is usually given in the tables. An additional choice concerns the evaporation source, which must be made from a material which, if possible, does not react with the evaporant or form compounds with it. The most important properties of frequently used substances are given in Tables 4 and 5.

For evaporation, pure materials are mostly used. If purity is to be preserved in the deposited film, the residual gases should be kept at sufficiently low pressure during evaporation. Additional sources of possible contamination exist at the beginning of the process: the increase in pressure caused by liberation of the gas adsorbed in both evaporant and evaporation source, and impurities arising from the evaporated layers of chemical compounds (e.g. oxides, sulphides) formed on the surface of the evaporant. To prevent condensation of these substances on the substrate, the latter is shielded by a shutter at the beginning of the evaporation.

*Table 4*

| Material | Atomic (molecular) weight | Density | Melting point (°C) | Region of the evaporation temperature* | Evaporation source** |
|---|---|---|---|---|---|
| Ag | 107.88 | 10.5 | 961 | 3 | Mo, W, Ta |
| Al | 26.98 | 2.7 | 660 | 3—4 | W, Ta |
| Au | 197.20 | 19.3 | 1 063 | 4 | W, Mo |
| Be | 9.02 | 1.9 | 1 284 | 4 | Ta, W, Mo |
| Bi | 209.00 | 9.78 | 271 | 2 | W, AO, Mo, Ta |
| C | 12.01 | 1.2 | 3 700 | 6 | C (arc) |
| Cr | 52.01 | 6.8—7.1 | 1 900 | 4 | W |
| Cu | 63.57 | 8.85—8.92 | 1 084 | 4 | W, Ta |
| Fe | 55.84 | 7.9 | 1 530 | 4 | W, AO |
| Ge | 72.60 | 5.35 | 958 | 4 | W, AO, C, E |
| In | 114.76 | 7.3 | 156 | 3 | W, Mo |
| Ni | 58.69 | 8.85 | 1 453 | 4 | W, C, E |
| Pt | 195.20 | 21.5 | 1 773 | 5 | W, C, E |
| Pb | 207.21 | 11.3 | 328 | 4 | Fe, Ni, W, Mo AO, E |
| Se | 78.96 | 4.5 | 220 | 1 | W, Mo |
| Si | 28.06 | 2.4 | 1 415 | 4 | C, E |
| Sn | 118.70 | 7.28 | 232 | 3—4 | Mo, AO |
| Ti | 47.90 | 4.5 | 1 727 | 5 | W, C, E |
| Zn | 65.38 | 7.13 | 420 | 2 | W, C, Ta, Mo, AO |
| Zr | 91.22 | 6.53 | 1 860 | 5—6 | C, E |
| Ni—Cr | — | 8.2 | — | 4—5 | W, Ta |
| SiO—Cr | — | | | 4—5 | W |

*, ** See footnote after Table 5

*Table 5*

| Material | Atomic (molecular) weight | Density | Melting point (°C) | Region of the evaporation temperature* | Evaporation source** | $\varepsilon_r$, Bulk value | n (Refractive index in the visible spectrum) | F, breakdown (MV/cm) |
|---|---|---|---|---|---|---|---|---|
| $Al_2O_3$ | 101.94 | 3.6 | 2 046 | 6 | E | 8.6 | 1.6—1.74 | |
| $CeO_2$ | 172.12 | 6.9 | 2 600 | 5—6 | W | | | |
| $MgO$ | 40.32 | 3.65 | 2 640 | 6 | E | | 1.736 | |
| $SiO_2$ | 60.09 | 2.1 | 1 500 | 5—6 | E | 3.7 | 1.46 | 3 |
| $SiO$ | 44.09 | 2.1 | – | 4 | Mo, W, Ta | | 1.9 | 2 |
| $ThO_2$ | 264.10 | 9.69 | 3 050 | 7 | E | | 2.20 | |
| $Ta_2O_5$ | 441.76 | 8.7 | 1 470 | 5 | Ta, W, E | 27 | – | 08 |
| $CdS$ | 144.46 | 4.8 | 1 750 | 3 | W, Mo, Ta | | 2.50 | |
| $ZnS$ | 97.44 | 3.9 | 1 900 | 3 | Mo, E | 8.3 | 2.3 | 02—1.5 |
| $CaF_2$ | 78.08 | 3.2 | 1 360 | 3 | Ta, W | | 1.43 | |
| $Na_3AlF_6$ | 209.95 | 2.9 | 1 000 | 3 | Mo | 6 | 1.33 | 1 |
| $LiF$ | 25.94 | 2.6 | 870 | 3 | Ta | 9,2 | 1.39 | |
| $MgF_2$ | 62.32 | 2.9—3.2 | 1 220 | 4 | Ta, AO, W | 5 | 1.38 | 2 |
| $NaF$ | 41.99 | 2.8 | 990 | 4 | Mo, Ta, W | | 1.32 | |
| $PbTe$ | 334.78 | 8.1 | 904 | 3 | Mo, Ta, AO | | – | |
| $CdTe$ | 240.0 | 6.2 | 1 041 | 3 | Mo, Ta | | 2.5 | |
| $NaCl$ | 801 | 2.2 | 801 | 3 | Ta, W, C | | 1.54 | |

* In Tables 4 and 5: (1) 100 to 400 (2) 400 to 800; (3) 800 to 1 200; (4) 1 200 to 1 600; (5) 1 600 to 2 100; (6) 2 100 to 2 800; (7) 2 800 to 3 500; all in °C

** W — Tungsten; Mo — Molybdenum; Ta — Tantalum; C — Graphite; AO — Aluminium Oxide (crucible); E — Electron-Beam Evaporation.

Similarly, evaporation of all substance from the source is not recommended because at the end of evaporation there is a danger of film contamination by the source material.

### 2.3.24 Evaporation Sources

It can be seen from the Tables 4 and 5 that the most commonly used materials for evaporation sources are the metals with a high melting point, i.e. W, Ta and Mo. They are used in the form of wires, foils or specially shaped

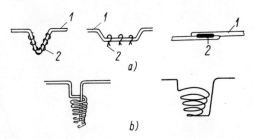

*Fig. 18:* Filament evaporation sources: (a) 1 — the wire is made from high-melting-point material directly heated by passage of current; 2 — material to be evaporated; (b) basket of high-melting-point wire.

boats. The evaporated material — a wire — is wound or hung in the form of a 'slider' on the wire of the source (Fig. 18a) or a basket is wound from tungsten wire, which is sometimes covered by a film of $Al_2O_3$ to prevent

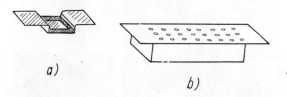

*a)*

*b)*

*Fig. 19:* Evaporation boats.

short-circuiting of turns by the metal evaporant placed in it (Fig. 18b). The shapes of the boats are shown in the illustrations. There exists a danger during evaporation of certain substances, as for example SiO, from the boat depicted in Fig. 19a that large particles will be ejected which would cause defects in the deposited film. To prevent this, boats with a perforated

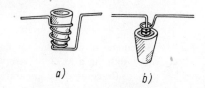

*Fig. 20:* Evaporation crucibles: (a) with outside, (b) with inside heating spiral.

*a)*

*b)*

cap (Fig. 19b) are employed or special systems are used in which the evaporated particles should be reflected from the heated plate before exit from the source. Some materials are evaporated from ceramic crucibles, which are heated either externally (Fig. 20a) or internally (Fig. 20b) by a spiral heater.

### 2.3.25 Special Evaporation Techniques

The methods described cannot be used for materials that have too high a melting point or decompose during the normal evaporation process.

For carbon, evaporation in an arc discharge between two graphite electrodes is used.

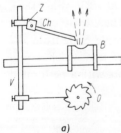

a)                                                          b)

*Fig. 21:* Schematic of apparatus for flash evaporation: (a) O — motor-driven gear wheel, V — vibrating arm, Z — reservoir, Ch — shaped chute, B — heated boat; (b) commercial apparatus (courtesy of Balzers Corp.).

In some applications, a technique known as a 'flash' is used. It is an instant evaporation of a small amount of material which is dropped onto the hot surface of the source. The technique is well suited for multicomponent compounds or mixtures, from which thin films can be prepared that have the same composition as the parent material.

The evaporant should be in the form of a powder which is dispensed from a reservoir in small quanta with the help of a mechanical or ultrasonic vibrator.

A very rapid evaporation of a given amount of metal can be achieved by the passage of a strong current pulse ($\approx 10^6$ A/cm$^2$) through a thin wire. The pulse is obtained by rapid discharge of a capacitor ($10 - 100\,\mu F$) charged to a sufficiently high voltage ($1 - 20$ kV). This is called exploding-wire evaporation. If the discharge energy is high enough and the duration of the pulse sufficiently short ($\sim 100\,\mu s$), it is possible to attain a deposition rate $\sim 10^5$ nm/s. Though the mechanism is not completely understood, it appears that at high temperatures (estimated at $10^6$ K) the wire is turned into a plasma arc from which particles are ejected, a considerable part in the form of ions. If the process proceeds too slowly, the filament does not disintegrate at once and individual, sizable droplets are formed instead of a homogeneous film.

Laser beam evaporation has come also into use recently. The laser source is situated outside the evaporation system, the beam penetrates through a window and is focused onto the evaporated material, which is usually in a fine-powder form. The light penetrates to the depth of only $\leq 100$ nm and the material evaporates therefore only at the surface. By this method it is possible to obtain a deposition rate of $10^5$ nm/s (energy 80 to 150 J; pulse duration of $2 - 4$ ms). The rate is actually higher since most of the evaporation takes place in a considerably shorter period, i.e., in the initial $10 - 100\,\mu s$. The temperature of the source is estimated to be $\approx 20\,000\,°K$ and emitted particles are mostly ionized (the process involved is apparently thermoionization).

All three methods just described have high deposition rates and the vapor particles arriving with a high speed at the substrate are mostly charged. As we shall see later, these circumstances influence the mechanism of film formation and so the films prepared by these methods may differ in their properties from those prepared by a normal evaporation.

Electron-bombardment is now the most commonly used evaporation technique. An electron beam of sufficient intensity is ejected from a cathode, $C$, accelerated and focused onto an evaporant material, which is heated at the site of incidence to the temperature required for evaporation. The method enables us to attain a very high temperature and to evaporate materials which would otherwise be evaporated with difficulty or cannot be evaporated at all. An additional advantage of the method is the prevention of contamination by the evaporation source material: the beam heats only the evaporant whereas the support holder is usually cooled. At the same time

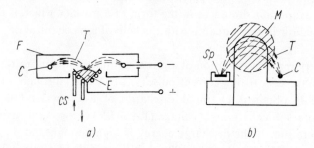

*Fig. 22:* Electron guns for electron-bombardment evaporation: (a) system with electrostatic focusing: C — heated cathode, E — evaporant (anode), F — focusing electrode, CS — water cooled spiral, T — electron trajectories; (b) system using magnetic focusing; C — cathode, Sp — specimen, M — magnetic field, T — electron trajectories.

no particles emitted from the heated cathode — the source of the electron beam — can reach the substrate. Some variants of this system, also called the electric gun, are shown in Fig. 22. Fig. 22a illustrates the system in which the electrons are focused onto a target electrostatically by means of a focusing electrode, F. The system in Fig. 22b employs a magnetic field, M, to bend the path of the electrons, so that the target cannot be contaminated

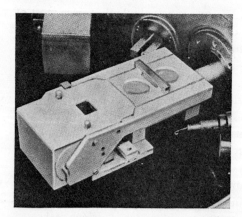

*Fig. 22 c:* A commercial electron gun with magnetic deflection (by permission of Balzers AG).

by particles evaporated from the cathode. The power supplies used in laboratory installations are in the range of kW to tens of kW.

The possibility of deflecting the electron beam by an electric or a magnetic field allows the use of the same gun for subsequent evaporation of various materials.

The accelerated electrons penetrate to a certain depth of the material (depending on their speed) and do not lose the main part of their energy before the end of their range. This means that most of the heat is liberated at a certain depth below the surface. It is therefore necessary to pre-degas the material well, so that the gases desorbed internally do not destroy the target.

### 2.3.26 Masking Techniques

For applications in electronics and microelectronics it is necessary to form the films in specific, sometimes complicated patterns. For this purpose a suitable mask is used to prevent vapor condensation on the areas which we desire to keep clean.

The mask is usually formed from a metal and is placed in the vapor stream as close as possible to the substrate on which the film is to be deposited. Resolution of the film shape is determined by the accuracy of the mask and its distance from the substrate. Larger masks are prepared mechanically or by chemical etching. When several masks have to be used to produce a multicomponent system, difficulties arise in precise adjustment of the masks. As it is usually desirable to prepare the whole thin-film system without interruption of the pump-down cycle, circular holders have been developed by which the masks *in vacuo* can be exchanged and the substrates placed above different evaporation sources. Such manipulation is not possible in the case of microelectronic elements where the dimensions involved lie in the range of microns. For such elements, 'in contact' masks are used which are deposited directly on the substrate.

There are several techniques for preparation of 'in contact' masks. The most widely used is the photolithographic one. The substrate is covered by a special emulsion which when exposed to uv light ceases to be soluble in an organic solvent due to polymerization. The unexposed parts remain soluble. The required pattern of the film is accurately drawn on a very large scale (e.g. 1.5 × 1.5 m) and is then reduced photographically to the required dimensions (e.g. 20 × 20 mm) and projected onto a given substrate. The unexposed parts are removed by a solvent and the film is deposited over the mask thus prepared. This method is combined with selective etching, where a suitable solvent removes a layer formed by one material while a layer from another material remains intact.

The method is best illustrated by an example as in Fig. 23, which shows the formation of a nichrome resistor with gold contacts. The use of photo-resist makes it possible to produce complex systems with great accuracy (e.g. 2.5 μm with an accuracy of ±1 μm).

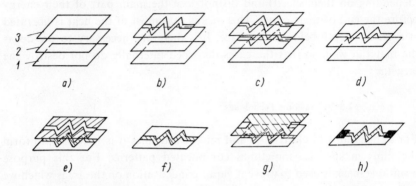

*Fig. 23:* Processing steps in the production of a nichrome resistor with gold contacts by using photo-resist: (a) glass substrate (1) is covered by evaporated Cu film (2) which is then sprayed with photo-resist (3); (b) photo-resist layer is illuminated through a master negative and the unexposed parts are removed (the photoresist is developed); (c) the parts of copper without the layer are etched away with a suitable agent; (d) polymerized photo-resist layer is removed with a solvent; (e) nichrome is evaporated onto the surface; (f) copper is etched away leaving only nichrome pattern; (g) photo-resist is redeposited and printed so as to enable contacts to be deposited; (h) gold is evaporated and photo-resist removed revealing film resistor with gold contacts

Recently electron beams have come into use in the technology of microelectronic circuits, either as machining devices for the cutting of accurate masks (the masks can be obtained also by an electroerosion) or for 'machining' of the film proper. In the latter case, the bombardment removes (evaporates) those parts which would otherwise be shielded by the mask. The difficulty in machining by an electron beam is that the material removed forms droplets on the edge of the kerf. The resolution with this method is much the same as that of the photolithographic one.

Greater accuracy can be achieved by using resists produced by impingement of electrons. For example, when certain organic silicates are bombarded by electrons a layer of silica is formed which can serve as an 'in-contact' mask. An accuracy of not less than 100 nm is attainable by this method. In order to take full advantage of this accuracy, it is necessary to use other etching methods than chemical ones. Etching by an ion beam (layers of Mo, Ta, N, Fe are etched by chlor-ions) or high-frequency etching is used.

Experiments employing a laser beam as a machining agent to replace the electron beam have been carried out. The formation of films of required pattern without the use of masks, in such a way that the film is deposited directly by a beam of ions, has also been tried. As with the electron beam method, this requires perfectly focused beams and makes heavy demands on electron optics. More detailed information about the formation and application of masks can be found in [18].

CHAPTER 3

# THIN FILM THICKNESS
# AND DEPOSITION RATE
# MEASUREMENT METHODS

Thickness is one of the most important thin-film parameters since it largely determines the properties of a film. On the other hand, almost all properties on thin films depend on the thickness and can be therefore used for the thickness measurement. From this fact follows a great diversity in methods of measurement. The very concept of thickness depends on the method of measurement selected or, more accurately, different methods of measurement may yield different results, i.e. different thickness for the same film.

A thin film is not normally perfectly smooth and has therefore different thicknesses at different places. If the method employed measures the mass spread over a unit area, from which the average thickness is calculated using a given mass density, then the thickness obtained is called the 'mass thickness'. In this sense, we speak about 'thickness' of extremely thin films (0.1 nm, a fraction of a monolayer) which have in fact the structure of isolated islands. By using other methods, e.g. optical ones, we generally obtain a different result. We shall return to this point later.

Some methods can only be used for finished films, while others enable us to monitor thickness during the actual process of film formation. Monitoring methods are very valuable because they allow the preparation of a thin film of selected thickness. Moreover, they can be used for measurement of deposition rate by measuring the growth of thickness over unit time. As we shall see later, the deposition rate is a significant parameter as it influences the morphology of the film and, therefore, its various properties. Techniques of controlling deposition rates — which have grown in importance with expansion of the industrial production of thin films — are based on these methods.

Methods of monitoring thickness can be divided into several groups, including balance, optical, electrical and other methods (e.g. those based on emission and absorption of radiation, chemical analysis, etc.).

When carrying out thickness monitoring during deposition, it is necessary to take into account the fact that the detecting element is not at the same place as the substrate. It is necessary, therefore, to make appropriate corrections, depending on the geometrical configuration, as has been noted in Section 2.3.1 in connection with the distribution of the deposited film thickness.

## 3.1 Balance Methods

### 3.1.1 Microbalance Method

This method is based on direct determination of the mass of film which is evaporated onto the substrate. The microbalance used for this purpose has to meet a number of special requirements. Above all, it must be sufficiently sensitive (of the order of $10^{-8}$ g/ m$^2$), mechanically rigid, easily degassable at elevated temperatures and have an aperiodic damping.

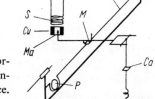

*Fig. 24:* Torsion balance developed by Mayer; T — torsion wire; M — mirror; S — solenoid; Cu — copper cylinder; Ma — magnet; P — spring, Ca — calibration device.

The perfect device from these points of view is the Mayer microbalance, which is shown in Fig. 24. The balance beam, in the form of a quartz wire, is fitted on one side with a light plate on which the material is evaporated, while on the other side a magnet is placed below a coil and surrounded by a copper cylinder to ensure the damping of vibrations by eddy current. A rider can be moved on the fork edge to obtain null balance and calibrate the system. The quartz torsion wire 120 mm long and 40 μm in diameter is fixed to a quartz frame by means of a quartz spring. A small mirror is mounted on the balance beam for precise optical determination of its position. Displacement brought about by thin film deposition is magnetically compensated by a proportional current in the solenoid. The method is sensitive to $10^{-8}$ g and makes it possible to measure thicknesses of a fraction of a monolayer. The system, made almost completely from quartz, can be degassed at temperatures up to 500 °C, so it is suitable for operation in an ultra-high vacuum.

There are a number of other microbalance systems which fulfill the above-mentioned requirements in varying degrees. In all cases we obtain a mass of film, $m$, per given area, $S$. The thickness $t$ is calculated by using the known mass density $\varrho$ of the substance. The value pertaining to the bulk material is taken for $\varrho$ even when the actual density of the thin film is usually lower. The thickness calculated from the relation

$$t = \frac{m}{S \cdot \varrho} \tag{3.1}$$

is therefore smaller than the real one.

### 3.1.2 Vibrating Quartz Method

This is one of the most frequently used methods. It is a method of dynamic weighing. The film is evaporated onto one electrode of quartz (accurately ground quartz plates with electrodes attached on opposite sides) connected to an oscillating circuit. The crystal of thickness $t$ has the fundamental resonance frequency

$$f = \frac{v_p}{2t} = \frac{N}{t} \tag{3.2}$$

where $v_p$ is the velocity of transverse elastic waves in the direction of the thickness $t$, and $N$ is the frequency constant. For an AT crystal cut (Fig. 25a) the constant is 1670 kHz mm. If a certain amount of material is now deposited on the crystal, the thickness increases by

$$dt = \frac{1}{\varrho_k S} dm \tag{3.3}$$

where $dm$ is the mass deposited (mass increase), $\varrho_k$ the quartz density, and $S$ the film area. The frequency changes by

$$df = -\frac{f^2}{N \cdot \varrho_k} \cdot \frac{dm}{S} \tag{3.4}$$

Equation (3.4) holds provided the film is thin enough. The frequency shift in such a case is equal to that arising from the change of crystal mass, because the elastic properties of the film do not yet play a role. The more the elastic properties of the film approach those of the quartz the greater the critical thickness above which the above-mentioned relationship does not hold. The properties required of the oscillator depend on the range of thickness to be measured. If $f_0$ is the initial frequency of the crystal at the beginning of deposition, then

$$C_{f_0} = \frac{f_0^2}{N \cdot \varrho_k} \tag{3.5}$$

is the mass-determination sensitivity. The sensitivity of the crystal is thus defined as the mass per unit area, $dm/S$, which effects a frequency change of 1 Hz. If the thickness to be measured is very thin, it is advisable to use high initial frequency.

The linear dependence of $df$ on $dm$ given by (3.4) has, however, a limited range of validity, because for higher shifts $df$ the quantity $f_0$ should be replaced by $(f_0 - df)^2 \approx f_0(f_0 - 2\,df)$. If, for example, the deviation from linearity is to be less than 1%, $df$ must not exceed 0.5% of $f_0$. The mass $(dm/S)_p$, corresponding to the maximum allowed frequency change $df_p$, decreases with increasing frequency.

The values of the variables are given for several fundamental frequencies in Table 6.

Basic Parameters Needed for Quartz Crystal Thickness Monitoring　　　　*Table 6*

| $f_0$ (Hz) | $C_{f_0}$ (Hz cm$^2$/g) | $dm/S$ (g/cm$^2$) | $df_p =$ $5 \times 10^{-3}\, f_0$(Hz) | $(dm/S)_p$ (g/cm$^2$) |
|---|---|---|---|---|
| $1.0 \times 10^6$ | $2.26 \times 10^6$ | $4.42 \times 10^{-7}$ | $5.0 \times 10^3$ | $2.23 \times 10^{-3}$ |
| $2.5 \times 10^6$ | $1.41 \times 10^7$ | $7.09 \times 10^{-8}$ | $1.25 \times 10^4$ | $8.83 \times 10^{-4}$ |
| $5.0 \times 10^6$ | $5.63 \times 10^7$ | $1.77 \times 10^{-8}$ | $2.5 \times 10^4$ | $4.41 \times 10^{-4}$ |
| $1.0 \times 10^7$ | $2.26 \times 10^8$ | $4.42 \times 10^{-9}$ | $5.0 \times 10^4$ | $2.23 \times 10^{-4}$ |

It is always necessary to choose a compromise between the sensitivity required and the maximum measurable film thickness. The range of thicknesses measured can be extended outside of the linear region with the help of calibration. If the mass load is too great, however, the crystal may start to vibrate in modes different from the original one (shear thickness vibration) or may cease to vibrate altogether.

It is necessary to consider additional factors which can affect the frequency shift and could thus cause errors in the measurement of thickness. The most important factor is the temperature of the crystal. The fundamental frequency of the crystal may change appreciably with the temperature. The temperature during the evaporation is affected by the radiant heat of the evaporation source. In addition, there is the heat liberated directly in the condensation of vapor on the crystal. The heat of condensation may significantly influence the local temperature. Both these sources of heat cannot, in principle, be eliminated, the second not even checked. Since the relationship between temperature and frequency varies for different crystals, the crystal with the smallest temperature dependence is always chosen. Such a cut is called AT-cut (at 14 MHz frequency) for which a temperature change of 1 °C corresponds to a mass change of $4 \cdot 10^{-9}$ g cm$^{-2}$ (i.e. to an average thickness of 0.04 nm if the density is 1 g cm$^{-3}$). The crystal must be cut along the defined crystallographic direction with an accuracy of a minute of arc if the optimal temperature dependence is to be achieved. To reduce heating as much as possible, the crystal is mounted in a water-cooled

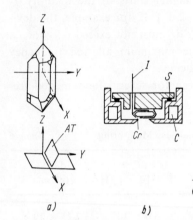

*Fig. 25:* (a) orientation of AT cut in quartz; (b) crystal holder: I — inlet bushing, S — seal, Cr — crystal, C — cooling.

holder and a shutter is stationed over it so only a small part of the crystal (e.g. 20%) is exposed to the stream of molecules. The construction of the holder is shown in Fig. 25b.

The change in frequency is usually measured by mixing the crystal frequency with a reference frequency and subsequent determination of their difference by a frequency counter. Time differentiation of the difference frequency gives the evaporation rate and it may be determined by an electronic differentiating circuit, or, if necessary, a digital system may be employed. By connecting the monitor to the evaporation source by means of a suitable feedback, it is possible to control the evaporation rate during evaporation and to terminate the evaporation with the help of an automatic shutter exactly at a specific frequency, i.e. at a selected thickness.

The crystal oscillator for measurement of thickness is usually sensitive to $10^{-9}$ g cm$^{-2}$ which, to give an example, amounts to 1/100 of a monolayer for iron. In circumstances where all possible extraneous effects are eliminated (e.g. temperature stabilized to within $\pm 0.01$ °C), a sensitivity of $10^{-12}$ g cm$^{-2}$ can be attained.

The apparatus can be constructed so as to permit degassing at higher temperatures and used in the ultra-high vacuum range.

It can be employed not only for monitoring of deposited films, but also for observation of film scouring by e.g. ion bombardment.

## 3.2 Electrical Methods

### 3.2.1 Electric Resistivity Measurement

Measurement of electric resistivity is one of the simplest methods for the determination of thickness of thin metal films. With decreasing thickness, however, the resistance increases at a greater rate than would be expected. As we shall see, this is partly due to the scattering on the film boundaries

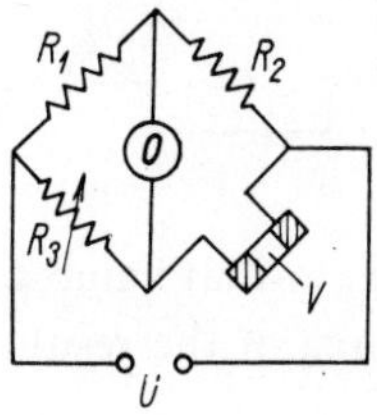

Fig. 26: Bridge method of film resistivity measurement: V — film; U — voltage; O — balance indicator; $R_1$, $R_2$, $R_3$ — resistors.

and partly due to the fact that the film structure differs from that of the bulk material and because the adsorbed and absorbed residual gases influence resistivity. In addition, conductivity changes for ultra-thin films, which are not continuous and exist in the form of islands and which behave in an entirely different way (see Sect. 6.2.2). Nevertheless, this method is applicable to a relatively extensive range of thickness, especially at higher deposition rates and low residual gas pressures.

The film is measured (between two contact strips) by means of a Wheatstone bridge (Fig. 26). By this method film resistivity ranging from fractions of an ohm to hundreds of megaohms can be measured. By using a relay-operated shutter the termination of evaporation at a specified resistivity can be ensured. With ordinary apparatus, the resistivity may be measured with a 1% accuracy; in special systems using a d.c. amplifier 0.01% accuracy has been achieved. The determination of thickness, however, depends on how well defined the relation of the resistivity to the thickness is. In practice, the accuracy is seldom better than $5^0/_0$.

### 3.2.2 Measurement of Capacitance

The thickness of dielectric films may be determined by measuring their capacitance. One possible arrangement is shown in Fig. 27. The evaporation of a film onto an electrode system effects a change in the capacitance measured.

An alternative approach is to measure the capacitance by covering the dielectric thin film on a conductive substrate by another (metal) film.

This method is often used for the measurement of thin oxide films formed on a metal surface. It may be applied, of course, only after the deposition is completed. Both methods require knowledge of the dielectric constant of the material, the value of the solid material being used for this. Uncertainty

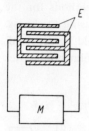

*Fig. 27:* Schematic of capacitance monitoring of film thickness: E — electrodes; M — capacitance bridge.

as to that value, together with any error in the determination of the surface area of the resultant capacitor, limits the accuracy of this method.

### 3.2.3 Measurement of Q-factor Change

If a thin metal film of thickness $t$ is placed a small distance $h$ from a coil of radius $r$ fed by alternating current, then some of the energy is lost by inducted eddy currents in the film. The Q-factor of the coil is thus changed together with the resonance frequency. A sufficiently high frequency must be used for ultra-thin films so that the skin-effect occurs at a depth comparable to the thickness (the penetration depth has to be greater).

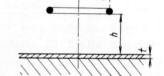

*Fig. 28:* Location of coil above film in Q-factor-change thickness monitoring.

    The frequencies employed are in the range of tens to hundreds of MHz and the measurable thicknesses are accordingly tens to hundreds of μm.

    The measurement is carried out as follows. The coil is made part of a bridge, the balance of which is disturbed by the interaction described above, and the imbalance results in a diagonal current which may be measured. If there is a need for measurements at very high frequencies, the bridge arrangement cannot be used. In such cases the coil is built into a resonance circuit whose damping and detuning is then measured.

The method has the advantage of being nondestructive and may also be exploited to control the evaporation process.

### 3.2.4 Ionization Methods

These methods are based on the measurement of ion current produced by ionization of the evaporant vapor. The stream of vapor is thus measured, i.e. the evaporation rate, from which the thickness of the film can be calculated, assuming that all particles impinging on the substrate are condensed there, i.e. the condensation coefficient equals unity.

The system most frequently used is basically an ionization gauge immersed in the stream of the evaporated material (Fig. 29). The electrons emitted from a heated cathode, $C$, are accelerated by a positive grid and ionize the molecules by colliding with them. The ions are captured on a collector which is negatively charged relative to the cathode and the ion current so produced is measured. If the vapor stream goes through an ionization chamber, it is ionized together with the residual gas. It is therefore necessary to separate the ions of the evaporant from the ions of the residual gases. Two techniques have been developed. One of them employs a second ionization gauge of the same construction which is placed so that the vapors do not pass through its ionization chamber and which consequently measures only the ionic flux originating from the residual gas. The difference between

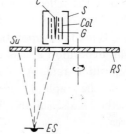

Fig. 29: Thickness monitor with ionization gauge: ES — evaporation source; Su — substrate; RS — rotary shutter; S — screening; C — cathode; G — grid; Col — collector.

the readings on the two gauges is then the measure of the number of particles evaporated. In the second technique, a vibrating or a rotating shutter modulates the vapor stream so that the ion current has a d.c. residual-gas component and a modulated one from the measured vapor stream. The latter is selectively amplified and recorded. This ion current is a linear function of the deposition rate, which represents the great value of the method.

The electrodes of the ionization vacuometer must be constructed to allow for heating to prevent condensation of the measured material on them (important especially in evaporation of dielectrics) or at least permit periodic cleaning.

It is possible to measure evaporation rates of $\sim 0.1$ nm/s with an accuracy of $1-5\%$, provided the measuring devices have an adequate sensitivity. Pre-selected thicknesses may be achieved with an accuracy within $\pm 1$ nm.

Apparatus of this type is well suited to serve as a gauge for automatization of evaporation systems and is frequently used in serial production of thin film systems.

# 3.3 Optical Methods

### 3.3.1 Method Based on Measurements of Light Absorption Coefficient

If light from a source of intensity $I_0$ passes through a thin film of a substance which absorbs the light, its intensity is reduced to a value $I$ (which can be measured by a suitable photocell) according to the equation

$$I = I_0 . (1 - R)^2 . \exp(-\alpha t) \tag{3.7}$$

where $t$ is the film thickness, $\alpha$ its absorption coefficient for a given radiation, and $R$ the reflectance of the air-film boundary. We can obviously utilize the equation for measurement of film thickness. The method is very simple and is used in evaporation of metals. It may be applied during the process of film deposition and is thus suited to control the process and monitor the deposition rate. At a constant deposition rate, and on semilogarithmic scale, the graph of transmitted light intensity vs. time is a straight line. The method is also suitable for checking the uniformity of thickness over a given area.

It should be noted, however, that only those substances that form continuous microcrystalline films of small thicknesses (e.g. $Ni-Fe$ alloy) behave in this way. Other substances, e.g. silver, behave somewhat differently; at small thicknesses (up to $\sim 30$ nm) the intensity of transmitted light decreases linearly with the thickness and only later is there an exponential decrease. Hence it is necessary to find out for each material whether equation (3.7) holds, or, if necessary, to determine the calibration curve.

### 3.3.2 Interference Methods

There is a whole number of optical methods for measurement of film thickness based on light interference of thin films. Visible light is an electromagnetic undulation with wavelengths ranging approximately from 400 to 800 nm. The interaction of two or more rays give rise to interference phenomena the intensity of the light is increased or reduced in certain directions. Interference in thin films may be explained with the help of Fig. 30.

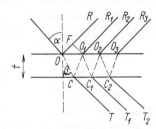

*Fig. 30:* Interference of light in a thin film.

A plane wave is impinging on a film of thickness $t$ at an angle $\alpha$. At the interferace, the wave is partially reflected and partially refreacted at an angle $\beta$. The refracted wave is again partially reflected and refracted at the lower interface and the process is repeated successively, as shown in the diagram. The initial reflected wave represented by a ray $OR$ interferes with waves denoted by $O_1R_1$, $O_2R_2$, etc., and wave $CT$ with waves $C_1T_1$, etc. For a transparent material, the reflected wave has usually much lower intensity than the refracted one and it is thus sufficient in a first approximation to consider the interference of only the first two adjacent waves.

The interference phenomenon depends on the difference in the optical paths of the interacting rays (The optical path $l$ is given as a product of the geometrical path and refractive index). Let us assume that the medium around the film is air $(n_{\mathrm{sub}} = n_{\mathrm{amb}} = 1)$ and that the film itself has refractive index $n$. The optical path difference of the $OR$ and $O_1R_1$ rays is

$$\Delta l = (OC + CO_1)\,n - OF = \frac{2tn}{\cos\beta} - 2 \cdot t \cdot \mathrm{tg}\,\beta \cdot \sin\alpha \qquad (3.8)$$

As the refractive index is given by

$$n = \frac{\sin\alpha}{\sin\beta} \qquad (3.9)$$

it is possible to transform the equation into the form

$$\Delta l = \frac{2t}{\cos\beta}\frac{\sin\alpha}{\sin\beta} - 2t \cdot \mathrm{tg}\,\beta \sin\alpha = 2tn\cos\beta = 2t \cdot \sqrt{(n^2 - \sin^2\alpha)} \qquad (3.10)$$

This equation holds whenever the phases of both waves are equal at the beginning. On reflection, however, the phase is, in some cases, reversed. It follows from both theory and experiment that the reversal occurs on reflection from the medium of higher optical density, i.e. of greater refractive index, whereas on reflection from the medium of lower optical density the phase does not change. This means that the reflection at point $O$ produces a change of phase by $\pi$, while the reflection at point $C$ has no effect (provided the refractive index of the film $n$ is greater than the indices of the substrate $n_{sub}$ and of the ambient atmosphere $n_{amb}$, $n_{sub} < n > n_{amb}$).

This reversal may be incorporated in the theory by expressing the effective path difference of the two interacting waves as

$$\Delta l_{ef} = \Delta l - \frac{\lambda_0}{2} = 2t \cdot \sqrt{(n^2 - \sin^2 \alpha)} - \frac{\lambda_0}{2} \qquad (3.11)$$

where $\lambda$ is the wavelength of the light *in vacuo*. If the effective difference is an integral multiple of the wavelength, the waves reinforce each other. Thus the condition for maximum interference is

$$\Delta l_{ef} = \Delta l - \frac{\lambda_0}{2} = 2k \frac{\lambda_0}{2} \qquad (3.12)$$

where $k$ is an integer, or alternatively,

$$\Delta l = (2k + 1) \cdot \frac{\lambda_0}{2} \qquad (3.13)$$

A minimum occurs, on the other hand, if the effective path difference is equal to an odd multiple of a half-wavelength, i.e.

$$\Delta l_{ef} = \Delta l - \frac{\lambda_0}{2} = (2k - 1) \frac{\lambda_0}{2}$$

or

$$\Delta l = 2k \frac{\lambda_0}{2} \qquad (3.13)$$

The number $k$ denotes the order of a maximum or minimum interference.

If the reflected intensity is not much lower than the refracted one, we have to consider the interference of further waves. The result expressed in the equations just presented is not, however, modified to any great extent. The positions of the maxima and the minima remain unchanged, but the intensity of the former will be greater. The interference phenomenon of the transmitted light is complementary to that of the reflected light. In our case, the phase reversal does not occur and the condition for a maximum will be thus given as

$$\Delta l = 2t \sqrt{(n^2 - \sin^2 \alpha)} = 2k \frac{\lambda_0}{2} \qquad (3.14)$$

and for a minimum as

$$\Delta l = 2t \sqrt{(n^2 - \sin^2 \alpha)} = (2k + 1) \frac{\lambda_0}{2} \qquad (3.15)$$

If we wish to observe the interference pattern we have to concentrate the rays by a converging lens, e.g. the lens of a human eye; the pattern may therefore be observed directly.

The equations are easily confirmed by experiment as far as reflected light is concerned. When using transmitted light and transparent thin film with not too large a refractive index ($n \leq 2 - 2.5$), the interference pattern is not observed in practice because the intensity of the transmitted (refracted) wave is much greater than that of the reflected one (the intensities of multiply reflected waves are still lower). Hence interference phenomena cause only small relative variations in the intensity that the eye cannot easily distinguish.

(a) *Measurement of thickness by use of interference colors.* In the simplest case, when light strikes a film perpendicularly the maxima occur (provided $n_{amb} < n < n_{sub}$) for reflected light for film thicknesses $\lambda_0/4$, $\frac{3}{4}\lambda_0$, etc., the minima for the thicknesses $\lambda_0/2$, $\lambda_0$, etc. The opposite holds for transmitted light. When white light is used, the components which are reinforced in reflected light are those for which the film thickness amounts to an odd multiple of the quarter-wavelength, the attenuated ones those for which the thickness is an even multiple. The film, therefore shows the color in reflected light corresponding to a combination of the enhanced components. When viewed by transmitted light the film displays the complementary color.

If we observe the film as it is formed by, e.g. evaporation, and exposed to white light, it assumes various colors; tables exist giving the thickness equivalent to an observed color. If the process continues, the sequence of colors (violet, blue, green-blue, green, yellow-green, yellow, orange, red, purple) recurs again in each order (i.e. for successive $k$). Different authors, however, describe the colors for particular thicknesses somewhat differently as subjective considerations enter here. The method, however, provides a good criterion when it is necessary to form a number of films of the same thickness. By observing the color changes, it is possible to stop the evaporation after a certain number of cycles and on appearance of the required hue.

A variant of this method which makes it possible to obtain objective readings rests on the reflection and transmission of light through a thin, weakly absorbing film deposited on a transparent substrate of a different refractive index. The intensities of both reflected and transmitted light vary periodically with film thickness, and therefore exhibit a whole series of maxima and minima. The intensities may be measured by a photometric technique. The periodic variations of the reflectance with thickness for films of various refractive indices deposited on glass of refractive index 1.5 are

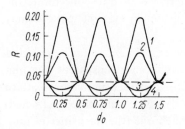

*Fig. 31:* Variation of reflectance $R$ with optical thickness (measured in units of wavelengths) for materials with different refractive indices: (1) $n = 2.0$; (2) $n = 1.75$; (3) $n = 1.4$; (4) $n = 1.2$ (the index of glass substrate is 1.5).

shown in Fig. 31. The medium above the film is air $(n_{amb} = 1)$. The relative variations of the reflection may be considerable (e.g. 21% maximum, 4% minimum for SiO on glass). The more the film absorbs, the less pronounced are the maxima.

The phenomenon may be measured objectively by means of a spectrometer. If the $k$th-order maximum occurs at wavelength $\lambda_1$ and the $(k + 1)$ th-order one at wavelength $\lambda_2$ at normal incidence then

$$2nt = k\lambda_1 = (k + 1) . \lambda_2$$

or

$$2nt = \frac{\lambda_1 \lambda_2}{\lambda_1 - \lambda_2} \tag{3.16}$$

where $n$ is the film refractive index (assumed to be the same for both wavelengths). If the refractive index is known, it is possible to determine the thickness $t$, and vice versa.

If the substrate of the film is absorbing, complicated effects of phase shifts arise on reflection.

(b) *Methods using interference fringes of equal thickness.* If the interference of monochromatic light is established on a thin film having the form of a wedge, the conditions (3.12) and (3.13) for interference maxima and minima will be fulfilled for certain thicknesses, hence alternate dark and light parallel fringes will be observed. For film with irregular distribution of thickness, the fringes will also have an irregular shape. The interference method, developed mainly by Tolansky, has become one of the standard methods for thin film measurement.

A diagram of the measurement system is shown in Fig. 32. The layer on which the interference takes place is formed by an air gap between two optical flats inclined by a small angle, one of which supports the measured film, which constitutes a kind of step on the surface. Both flats are metal-

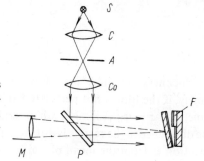

Fig. 32: Tolansky's method of thickness measurement: S — light source; C — condensor; A — aperture; Co — collimator; P — plate with semi-transparent layer; M — microscope; F — film to be measured.

coated with the same substance. A parallel beam of rays is directed from a monochromatic source at an angle of 45° onto a semi-transparent mirror and is partially reflected to the interference system. In the system, interference fringes of equal thickness arise separated by $\lambda/2$ when viewed with a low-power microscope. If both reflecting planes have a sufficiently high reflection coefficient and their spacing is small, then the interference fringes are very sharp (the thickness of dark fringes on a white background may be reduced to as little as $\lambda/100$). In the arrangement the fringes run parallel to the contact

Fig. 33: Fringe displacement at the film step in Tolamsky's method.

edge of the flats. At the site of the film step the fringes are displaced (Fig. 33). If $L$ is the fringe spacing, $\Delta L$ the displacement of the fringes, then the film thickness is given by

$$t = \frac{\Delta L}{L} \cdot \frac{\lambda}{2}$$

(3.17)

where $\lambda$ is the wavelength of the monochromatic light. It is necessary to coat the plate with the film as well as the opposite one with a layer possessing the same high reflection coefficient in order that phase changes on reflection will be the same at both sides of the air wedge. If the process is executed carefully, it is possible to measure even ultra-thin films with adequate accuracy (e.g. a film of 5 nm with an accuracy of $\pm 0.8$ nm). The value of

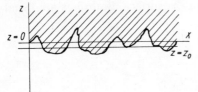

*Fig. 34:* Reflection of light on a rough plane.

the thickness obtained may however, be influenced by another factor, namely, if the film is not perfectly flat reflection takes place at an apparent plane, e.g. $z = z_0$ (Fig. 34), which is distinct from the mean plane $z = 0$. The difference in the results is directly proportional to the square of the mean deviation $\bar{\varepsilon}^2$ from the mean plane. If, for example, a silver layer evaporated on the measured film is smoother than the substrate plate, the error in the determination of thickness will be given by

$$\Delta t = \frac{2\pi\varkappa}{\lambda}\left(\bar{\varepsilon}^2_{\text{film}} - \bar{\varepsilon}^2_{\text{sub}}\right) \tag{3.18}$$

$\varkappa$ is the imaginary part of the complex refractive index of the layer. The error will arise in all interference methods and, as can be seen in Table 7, may amount to a large value for surface irregularities of an order of 10 nm.

Error in Thickness Measurement  *Table 7*
Caused by Surface Roughness

| $\sqrt{\varepsilon^2}$ (nm) | 5.0 | 10.0 | 15.0 | 20.0 | 25.0 |
|---|---|---|---|---|---|
| $\Delta t$ (nm) | 0.8 | 3.5 | 8.0 | 14.0 | 22.0 |

(c) *Interference methods using the interference microscope* (Fig. 35). This utilizes the interference of only two beams produced by a prism (in contrast to the multiple beam interference used in the previous method). Light from a monochromatic source $S$ is divided by the prism into two beams. One of them proceeds to the flat mirror from which it is reflected, interfering with the second one reflected from the substrate supporting the film. The resulting interference pattern is viewed with a microscope. In this case, the fringes are not so narrow and accuracy is less. It is still possible, however, to measure thicknesses of about 10 nm to an accuracy of about 2 nm. A considerable advantage of the method is that no mechanical damage is done to the film by contact with the second optical flat.

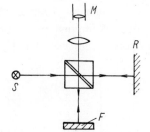

*Fig. 35:* Scheme of interference microscope: S — mono-chromatic light source; R — flat reference plane (mirror); F — film; M — microscope.

A number of other interference methods has been developed some of which enable the measurement of a thickness down to 1.5 nm with an accuracy of ±0.5 nm.

For thicknesses ranging from about 25 nm to 100 nm, interference of monochromatic X-rays may also be used. Recently a new method has been suggested that uses a laser as the light source and is based on the principle of holographic interferometry (see *J. Vac. Sci. Tech.*, **9** (1972), 1080).

### 3.3.3 Polarimetric (Ellipsometric) Method

A thin film (tenths of nm) partially or wholly transparent on a polished metal plate affects the ellipticity of reflected light. The polarimetric or ellipsometric method is based on measuring the ratio of the amplitude of the reflected light polarized along the plane of incidence and that polarized perpendicularly and their phase difference at relatively large angles of incidence. By this method it is possible to determine the thickness, or optical constants, of thin non-absorbing and absorbing homogeneous isotropic films on both non-absorbing and absorbing substrates. The thickness may be calculated from the data with the help of relatively complicated mathematical equations derived from the electromagnetic theory of light. The method has found a wide use only recently, the complicated calculations being carried out by computers.

For the examination of ultrathin transparent films deposited on a metal surface (e.g. films of oxides arising spontaneously on the surfaces of some metals such as Al, Ta, etc.), this is the only practicable method. It is, however, very laborious and not therefore used for *in-situ* monitoring of thickness during film formation *in vacuo*.

## 3.4 Deposition Rate Monitoring Using Transfer of Momentum

The property employed for monitoring in this case is the momentum of particles impinging on a movable element whose displacement is recorded.

Such a system is illustrated in Fig. 36. Particles from evaporation source $S$ impinge onto one half of a cylinder suspended on a torsion wire $T$, whereas the second half of the cylinder is shielded by a shutter $Sh$. The

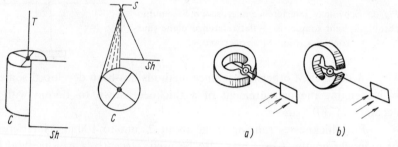

*Fig. 36:* Deposition rate monitoring by means of momentum transfer to a rotating cylinder: $S$ — source; $C$ — cylinder $T$ — torsion wire; $Sh$ — shield.

*Fig. 37:* Using Deprezs system for the measurement of (a) deposition rate; (b) film thickness.

momentum transferred to the rotor by a molecular beam causes angular displacement of the cylinder, from which, using known parameters of the system, the flow of the impinging particles may be calculated. If we wish to calculate the thickness of the evaporated film from the measured data, we have to know the condensation coefficient of the substance on the given surface, which depends also on temperature.

If the condensation coefficient is unknown but the thickness is obtained by some other method (e.g. balance or optical one), we can use the method for the determination of this coefficient. Another variant of the method is illustrated in Fig. 37. The molecular beam impinges on a plate connected with a rotating system of the Deprezs measuring device. In Fig. 37a, the weight of growing film is compensated by the rigidity of the system and the rotation is caused only by the momentum of impinging particles. The angular displacement may be compensated by an electric current in the coil winding. The current is thus a direct measure of the stream intensity of the particles. The method is in fact a null-balance method and is hence very sensitive.

In the arrangement of Fig. 37b, it is the momentum effect which is compensated by the rigidity of the system and the displacement is caused only by the static weight of the deposited film. This modification is a balance method of thickness measurement and therefore belongs to those described in Section 3.1.1.

## 3.5 Special Thickness Monitoring Methods

### 3.5.1 Stylus Method

The apparatus for this method (known commercially under the names Talysurf or Dektak) comprises a fine diamond tip with a radius of 0.7 to 2 μm which is pressed onto the surface with a pressure of 500 Kp/cm²

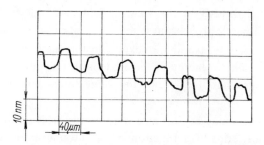

*Fig. 38:* Record of thickness obtained by the Stylus method.

(corresponding to a tip mass of only 0.1 g) and moves uniformly across it. The vertical movement of the tip, caused by irregularities of the surface, is converted to an electrical signal which is then amplified and recorded. A record of the profile of metal strips on an insulating substrate is shown in Fig. 38.

It is claimed that the minimal thickness difference recorded by this method is 2.5 nm with a precision of ±2% and that the detriment to the surface caused by the tip is minimal.

The method enables the distribution of thickness over the surface and the surface structure to be determined promptly and with relative accuracy. On the other hand, it does not record narrow cracks and crevices (the tip-surface contact has a comparatively large area). In comparison with the interference method it possesses the advantage that it is not necessary to coat the sample surface with an additional layer. The results obtained by the Stylus method are in good agreement with interference measurements, which testifies to their considerable reliability.

## 3.5.2 Radiation-absorption and Radiation-emission Methods

In most cases radiation is absorbed according to the exponential law (3.7). The exponential dependence can be used for thickness determination.

(*a*) The absorption of $\alpha$- or $\beta$-rays emitted from radioactive sources has been used for measurement of thicknesses ranging from several nm to several mm. In addition to $\beta$-rays a beam of electrons from an electron gun may be utilized. Thickness can be measured by use of the law of absorption only if the Bragg reflection does not occur (i.e. diffraction on crystal lattice). This condition is fulfilled to various degrees by materials of different crystal structures. (For example, microcrystalline films of Ni, Mn and Al behave well in accordance with the exponential law.) In this way it is possible to measure even ultrathin films with great precision (e.g. Ni films of 2 nm thickness with 5% precision). The method is particularly useful when the film is deposited directly in the viewing field of an electron microscope, where the required electron beam is available (see also Chapter 4).

The maximum measurable thickness in this method is determined by the maximum available energy of the electron beam. For energies of about 100 keV it is several hundreds of nm (this value differs, however, for different materials).

(*b*) Thickness may be also measured on the basis of a back scattering of rays. The scattering is a function of the atomic number, density and thickness of the sample. $\beta$-rays from radioactive sources $C^{14}$ (0.16 MeV) or $Pm^{147}$ (0.22 MeV) are employed and the scattered radiation is registered by a Geiger-Müller counter.

(*c*) The intensity of emitted X-rays, which may be excited in various ways, may be used as a measure of film thickness. The number of quanta emitted is proportional to the number of atoms capable of emission and hence proportional to the mass thickness of the measured film. The primary excitant of emitted radiation may be either 'white' X-rays (i.e. radiation with a continuous spectrum of frequencies) or a beam of electrons with sufficient energy. X-rays may also be excited by bombardment by fast protons (100 keV).

For thin films, the radiation of the film material itself is used and its intensity measured; in some cases the rays are excited in the substrate and then their attenuation in the film is measured.

The lower limit of thickness measurable is determined by the sensitivity of detection and is about 1 nm. With increasing film thickness the intensity of emitted X-rays approaches that excited in bulk material which is determined by the penetration depth of the primary excitant (e.g. an electron beam). For measurement *in vacuo*, the method is applicable for materials from atomic number 12 (Mg), and for measurement in air for materials from atomic number 22 (Ti). The line $K_\alpha$ is usually used, since it is the most intense in the fluorescence spectrum.

The method also permits the percentage composition of alloy films to be determined.

(*d*) If a certain known percentage of a radioactive substance is added to the evaporated material, the radioactivity of the evaporated film may be measured and the evaporated amount derived from pulse counter readings and the mass film thickness determined. The sensitivity of the method depends on the percentage of the added radioactive substance and it may even be used for the measurement of monolayers.

### 3.5.3 Work-function Change Method

In some special cases, the thickness of ultrathin films may be determined by measuring the work function. This method is chiefly employed for measurements of ultrathin films (mostly thinner than a monolayer) of strongly electropositive metals such as Cs or Ba on the surface of materials with a larger work function. With increasing coverage of the surface by the film, the work function decreases, and the extent of the coverage, i.e. the corresponding average film thickness, may be determined with considerable accuracy.

# CHAPTER 4

# MECHANISM OF FILM FORMATION

## 4.1 Formation Stages of Thin Films

It has been found by observation of films evaporated directly in the viewing field of an electron microscope that film growth may be divided into certain stages. These are as follows:

(1) Nucleation, during which small nuclei are formed that are statistically distributed (with some exceptions) over the substrate surface.

(2) Growth of the nuclei and formation of larger islands, which often have the shape of small crystals (crystallites).

(3) Coalescence of the islands (crystallites) and formation of a more or less connected network containing empty channels.

(4) Filling of the channels.

The process is schematically shown in Fig. 39. (It should be noted that Fig. 39a shows nuclei already formed because at the birth stage the

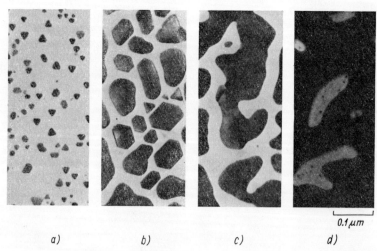

0.1 μm

a)          b)          c)          d)

*Fig. 39:* Process of formation of Ag film on MoS$_2$ (Pashley [26]).

nuclei dimensions are usually beyond the resolving power of an electron microscope.)

It is important to note that after a certain concentration of nuclei is reached additional impinging particles do not form further nuclei but adhere to the existing ones or to the islands formed already. As we shall see later, the nucleation process and the growth and coalescence of separate islands have a fundamental importance for the formation of the film structure, i.e. the size of crystallites, their orientation, etc. We shall therefore deal with these stages in more detail.

## 4.2 Nucleation

The particles which have been evaporated from the evaporation source and have reached the substrate, on which a thin film is to be deposited, generally lose part of their energy on impingement. They are attracted to the surface by forces mostly of dipole or quadrupole character and so they become, at least for a certain time, adsorbed on the surface (adatoms).

The energy loss of a particle which has impinged on the surface and then left it is characterized by the accommodation coefficient $\alpha$ defined as follows.

$$\alpha = \frac{T_c - T_v}{T_c - T_s} \tag{4.1}$$

where $T_c$ is the temperature corresponding to the energy of the incident particle (and determined in the main by the temperature of the evaporation source), $T_v$ is the temperature of the emitted particle and $T_s$ the substrate temperature. The coefficient values are in the range from 0 to 1. The zero value pertains to the case of elastic reflection (without energy loss), the unity value to total accommodation, when the particle loses all its 'excessive' energy and its energy state is then fully determined by the substrate temperature. The theoretical explanation has been elaborated for the simplified case of a particle indicent on a one-dimensional chain of connected particles. It has been found that the sticking coefficient is equal in practice to unity if the energy of the incident particles is less than or equal to about 25 times the desorption energy on the substrate. Since the desorption energy is usually about 1 to 4 eV, the upper limit corresponds to temperatures of the impinging particles in the region of $10^5$ K, which far exceed temperatures actually used in normal evaporation. It is therefore highly probable that most particles will be physically adsorbed. (When passing from the linear model to the spatial one, the probability, though decreased, would still remain high.) The adsorbed particles stay on the surface for a certain time $\tau_s$ given as

$$\tau_s = \frac{1}{\nu} \exp \frac{Q_{des}}{kT} \qquad (4.2)$$

where $\nu$ is the surface-vibrational frequency of the adatoms, $k$ is the Boltzmann constant, $Q_{des}$ the desorption heat of the particle on the given substrate and $T$ the 'temperature' of the particle which is generally between that of the evaporation source temperature and that of the substrate. The particle which has not been fully accommodated has retained a certain 'excess' energy. Due to this energy and the thermal energy from the substrate, the particle moves over the surface. The movement is called migration or surface-diffusion. During its stay on the surface the particle may be chemically adsorbed (chemisorption). The resulting state is of much higher desorption energy and the particle is then rarely re-evaporated (the time of stay, $\tau_s$, is very long). Besides this, it may happen that the particle will meet another one in the course of its surface-diffusion and form a pair with it which has a much lower re-evaporation probability than a single particle (desorption energy increases by the dissociation energy of the formed pair) and thus the conditions for condensation will be prepared.

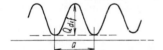

*Fig. 40:* Potential relief of the surface of a solid.

The condensation coefficient, which gives the ratio of the number of condensed atoms to the total number of impinging atoms, is important in thin film formation. The particle on the surface will move by means of the surface-diffusion to a certain mean distance $\overline{X}$ from the point of incidence. The distance is given as

$$\overline{X} = \left(2D_d\tau_s\right)^{1/2} = \left(2D_d\right)^{1/2} \nu^{-1/2} \exp\left(\frac{Q_{des}}{2kT}\right) \qquad (4.3)$$

where $D_d$ is the surface-diffusion coefficient. The surface may be represented as in Fig. 40. The binding energy is not the same over the whole surface and the adsorbed particle tends always to occupy the state of minimal energy. Thus it is always localized in some adsorption position (i.e. in some 'valley' of the relief) and to pass into an adjacent position it must overcome a certain potential barrier — the activation energy for a surface-diffusion jump. The surface-diffusion coefficient is related to this activation energy by

$$D_d = a^2 v \exp\left(-\frac{Q_{dif}}{kT}\right) \qquad (4.4)$$

Hence we may express the mean distance $\overline{X}$ as

$$\overline{X} = \sqrt{2} \cdot a \exp\left(\frac{Q_{des} - Q_{dif}}{2kT}\right) \qquad (4.5)$$

Several values of the magnitude of these quantities are listed in Table 8.

Values for the Desorption Energy $Q_{des}$     *Table 8*
and the Activation Energy for Surface
Diffusion $Q_{dif}$

| | $Q_{des}$ (eV) | $Q_{dif}$ (eV) |
|---|---|---|
| Ba on W | 3.8 | 0.65 |
| Cs on W | 2.8 | 0.61 |
| Al on mica | 0.9 | — |
| W on W | 5.83 | 1.21 |
| Hg on Hg | — | 0.048 |

If two adjoining nuclei formed by several particles are so near to each other that the regions (roughly of diameter $\overline{X}$) from which other particles can diffuse to them overlap, then the formation of additional nuclei will be practically stopped since all other particles will join existing islands. This implies that the concentration of nucleation centers may be determined from equation (4.5).

To ensure the formation of condensation nuclei, the evaporation rate must be sufficiently high, otherwise a particle migrating over the surface might re-evaporate before meeting another particle. This may be quantitatively expressed as follows. The concentration $n_1$ of individual particles on the surface is, under the assumption that the impinging flux is $N\downarrow$, given by

$$n_1 = N\downarrow \tau_s \left[1 - \exp\left(-\frac{t}{\tau_s}\right)\right] \qquad (4.6)$$

The $n_1$ is constant and equals $N\downarrow \cdot \tau_s$ for a sufficiently long duration of the adsorption $(t \rightarrow \infty)$. This means that in a stationary state the impinging current $N\downarrow$ is equal to the flow $N\uparrow$ of re-evaporated particles

$$N\downarrow = \frac{C \cdot p}{\sqrt{(2\pi m k T_{\mathrm{v}})}} = \frac{n_1}{\tau_{\mathrm{s}}} = n_1 \cdot v \cdot \exp\left(-\frac{Q_{\mathrm{des}}}{k \cdot T}\right) \qquad (4.7)$$

where $p$ is the vapor pressure corresponding to the temperature of the evaporation source $T_{\mathrm{v}}$, $C$ is a constant depending on the geometric configuration of the source and the substrate, and $m$ is the mass of the particle. At very low evaporation rates, the $n_1$ will thus be very small and the probability of the formation of condensation nuclei will be negligible. The best conditions for their formation and additional film growth will exist at high evaporation rates. During formation and growth of nuclei the process will, of course, not be stationary; the impingement flow is higher than the re-evaporation flow.

The ratio of the impinging flow to the re-evaporation flow $N\downarrow/N\uparrow$ is called supersaturation and it is an important parameter for thin film condensation. The re-evaporation flow is determined by the equilibrium pressure of the evaporant vapor at the temperature of the substrate (e.g. for Ag at 300 °K the pressure is $10^{-40}$ torr), whereas the impingement flow corresponds to a given evaporation rate (for Ag at 0.1 nm/s evaporation rate, i.e. approximately 1 atom Ag on 1 substrate atom per second, it corresponds to a pressure of $\sim 10^{-6}$ torr).

The conditions under which the condensation begins depend in the main on the ratio of two quantities: the desorption energy characterizing the binding of impinging atoms on the substrate, and sublimation heat, $Q_{\mathrm{s}}$, characterizing mutual binding of the condensing atoms.

($a$) If $Q_{\mathrm{des}} \ll Q_{\mathrm{s}}$, the condensation occurs without supersaturation ($P/P_{\mathrm{c}}$ may be less than 1) and the coverage is high.

($b$) If $Q_{\mathrm{des}} \approx Q_{\mathrm{s}}$, the condensation occurs at a moderate level of supersaturation. This is the region which is satisfactorily explained by the classical theory of nucleation based on thermodynamic concepts (capillarity theory).

($c$) If $Q_{\mathrm{des}} \gg Q_{\mathrm{s}}$, only very small coverage is achieved under normal conditions and a high supersaturation must be used to effect the condensation. The thermodynamic theory of heterogeneous nucleation is hardly applicable here and it is necessary to use the atomistic theory.

### 4.2.1 Capillarity Theory of Nucleation

The theory is based on thermodynamic concepts and stems from the Langmuir-Frenkel theory of condensation.

Let us assume that the condensation nucleus has a spherical shape of radius $r$. During its growth by joining of additional particles (mostly by surface-diffusion) the nucleus energy consisting of the surface and volume components will change. Gibbs's free energy (free enthalpy or Gibbs's potential) $\Delta G_0$ is

$$\Delta G_0 = 4\pi r^2 \sigma_{cv} + \tfrac{4}{3}\pi r^3 \, \Delta G_v \tag{4.8}$$

where $\sigma_{cv}$ is the condensate-vapor interfacial free energy and $\Delta G_v$ is the free-energy difference per unit volume of the phase of molecular volume $V$, condensed from the supersaturated vapor with the equivalent pressure, $p$, to the state with the equilibrium pressure $p_e$, i.e.

$$\Delta G_v = -\frac{kT}{V} \ln \frac{p}{p_e} \tag{4.9}$$

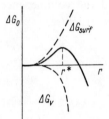

Fig. 41: The dependence of $\Delta G_0$ on nucleus radius.

Fig. 41 shows the dependence of $\Delta G_0$ on the radius of the nucleus. It can be seen that the $r$-dependence possesses a maximum at a certain critical radius $r^*$ which can be calculated from the condition

$$\frac{d(\Delta G_0)}{dr} = 0 \tag{4.10}$$

The $r^*$ value (i.e., radius of the critical nucleus) is therefore

$$r^* = -\frac{2\sigma_{cv}}{\Delta G_v} \doteq \frac{2\sigma_{cv}V}{kT \cdot \ln \dfrac{p}{p_e}} \tag{4.11}$$

If the radius of the nucleus is smaller than $r^*$, the nucleus is unstable and there is a high probability that it will disintegrate (the nucleus tends to occupy the lowest energy state which in this case requires disintegration into individual isolated atoms). If the radius is greater than $r^*$, the energy will decrease with increasing radius so the aggregates will grow until a continuous film is established. We can understand from this how important for film formation is the ratio $p/p_e$, which is called the supersaturation coefficient.

The nuclei do not have, however, spherical shape, but rather the shape of a spherical cap with a contact angle $\vartheta$ determined by the equilibrium of the surface forces (Fig. 42). Denoting the surface free energies by $\sigma$, with subscripts $s$, $c$, and $v$ referring to the substrate, condensation nucleus and vapor respectively, we obtain

$$\sigma_{cv} \cos \vartheta = \sigma_{sv} - \sigma_{sc} \tag{4.12}$$

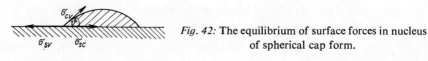

Fig. 42: The equilibrium of surface forces in nucleus of spherical cap form.

In a manner analogous to that used in equation (4.8), we may also write down Gibbs's free energy in the form of the sum of the surface energy $\Delta G_1$ and the volume energy $\Delta G_2$, which now assume a somewhat more complex structure.

$$\Delta G_1 = \pi r^2 \sin^2 \vartheta (\sigma_{sc} - \sigma_{sv}) + 4\pi r^2 \phi_1(\vartheta) \cdot \sigma_{cv} \tag{4.13}$$

$$\Delta G_2 = \frac{4\pi}{3} r^3 \phi_2(\vartheta) \cdot \Delta G_v \tag{4.14}$$

where $\phi_1(\vartheta)$ and $\phi_2(\vartheta)$ are geometrical factors and $\Delta G_v$ has again the form of (4.9). In the refined theory, still another term $\Delta G_3$ is added which represents the entropy of the distribution of the nuclei among $n_0$ possible positions on the surface of a crystal lattice.

$$\Delta G_3 = - kT \ln \left( \frac{n_0}{n_1} \right) \tag{4.15}$$

The sum of these three terms again reaches a maximum at a certain value $r^*$ and the corresponding critical energy $\Delta G^*$ of the nucleus formation; these are given by

$$r^* = - \frac{2\sigma_{cv}}{\Delta G_v} \tag{4.16a}$$

$$\Delta G^* = - kT \ln \frac{n_0}{n_1} + \frac{16\pi}{3} \frac{\sigma_{cv}^3}{\Delta G_v^2} \cdot \phi_3(\vartheta) \tag{4.16b}$$

where $\phi_3$ is again a geometric factor given by

$$\phi_3(\vartheta) = \tfrac{1}{4}(2 + \cos \vartheta)(1 - \cos \vartheta)^2 \tag{4.17}$$

Its dependence on $\vartheta$ is plotted in Fig. 43.

For complete wetting of the substrate by the condensate when $\vartheta = 0$, only the first term remains in (4.16b); $\Delta G^*$ is negative, which makes the situation favorable for the nucleation. In the case of zero wetting, $\vartheta = 180°$, $\phi_3 = 1$ and (4.16b) is transformed into the formula for homogeneous nucleation. This corresponds to the situation when the substrate has no catalytic effect on nucleation.

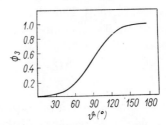

*Fig. 43:* The dependence of $\phi_3$ on $\vartheta$

The quantity $\Delta G^*$ may be affected by various factors. For certain values of $\vartheta$, $\Delta G^*$ may be, for example, smaller at the steps of a crystal lattice. This is the basis of the surface-decoration effect by means of which it is possible to make the steps visible by coating them with an ultrathin layer of suitable metal which condenses only on these steps (an example of such a decorated surface is shown in Fig. 16). Impurities on the surface may decrease or increase $\Delta G^*$, depending on their nature. Electrostatic charges on the surface reduce $\Delta G^*$ and thus facilitate condensation.

The nucleation rate $J$ is very important; it is proportional to the concentration of the critical nuclei $N^*$,

$$N^* = n_0 \cdot \exp\left(-\frac{\Delta G^*}{kT}\right) \qquad (4.18)$$

where $n_0$ is the density of adsorption sites, and $\Gamma$ — the rate at which molecules join the critical nucleus by surface-diffusion. (It follows from a comparison of the relative fractions of the particles joining the nuclei by surface-diffusion and by direct impingement from the vapor phase that the latter mechanism may by neglected.) We have

$$J = Z \cdot 2\pi r^* \Gamma N^* \sin \vartheta \qquad (4.19)$$

$Z$ is Zeldovich's constant and represents the departure of the real state from equilibrium and is about $10^{-2}$, $2\pi r^* \sin \vartheta$ is the periphery of the critical nucleus. For the rate $\Gamma$ we obtain from the diffusion process the relation

$$\Gamma = n_1 \cdot a \cdot v \cdot \exp\left(-\frac{Q_{\text{dif}}}{kT}\right) \qquad (4.20)$$

where $n_1$ is the concentration of adsorbed atoms (see (4.6)). By substitution we obtain the final expression (by using (4.7)):

$$J = Z \, 2\pi r^* n_0 a \sin \vartheta \, \frac{p}{\sqrt{(2\pi m k T)}} \exp\left(\frac{Q_{des} - Q_{dif} - \Delta G^*}{kT}\right) \quad (4.21)$$

(For the meaning of $a$ see Fig. 40). In this expression $\Delta G_v$ is incorporated into $\Delta G^*$ (see (4.16b)). Thus the nucleation rate is largely dependent on the supersaturation. The quantity $\Delta G_v$ has the critical value

$$\Delta G_{vcrit} = -\frac{kT}{V} \ln\left(\frac{p}{p_e}\right)_{crit} \quad (4.22)$$

which corresponds to critical supersaturation. Its significance is obvious from Fig. 44. The dependence of the nucleation rate on supersaturation is very strong: at a supersaturation lower than the critical one $J$ is practically

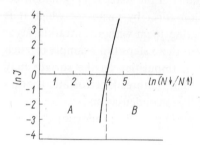

Fig. 44: Nucleation rate as a function of super saturation: A — region where no film formation occurs; B — region of film formation.

zero, whereas at supersaturations higher than the critical one $J$ increases very rapidly $(J \to \infty)$. We speak about the 'onset' of condensation when the nucleation rate reaches about 1 nucleus per second on 1 cm².

If the critical nucleus consists of at least two atoms and the free energy of its formation from vapor is positive, a certain energy barrier exists which prohibits the formation of a continuous film; the island structure appears. If the barrier is high ($\Delta G^*$ is great), the radius of the critical nucleus is large and so a relatively small number of large aggregates is formed. On the other hand, if the barrier is low (small $\Delta G^*$), a great number of small aggregates is formed and the film becomes continuous even at a relatively small thickness. It is therefore clear that factors will substantially affect the final film structure.

In the case of silver deposited on glass, we obtain $r^* \approx 0.46$ nm and a supersaturation of $10^{34}$; for tungsten under similar conditions $r^* \approx 0.13$ nm, the supersaturation is $10^{106}$, and the film is very soon continuous.

Supersaturation depends at a given temperature on evaporation heat, which in turn is related to the boiling point $T_b$ by the Trouton relation

$$\Delta L_{vap} = \frac{\Delta H_{vap}}{T_b} \qquad (4.23)$$

where $\Delta H_{vap}$ is the enthalpy change. Metals with high boiling point have high values of supersaturation, small-sized critical nuclei and they easily form continuous films. The size of the critical nucleus diminishes considerably whenever there is a strong adhesion between the substrate and evaporated film. The substitution of the theoretical values for the condensation of metal on metal may sometimes result in a negative critical radius. This is not, of course, possible, but it indicates that the critical nucleus contains less than two atoms and that an energy barrier does not exist (no island are formed).

The capillarity theory has enabled us to understand certain basic laws governing thin film formation and has provided a qualitatively correct idea about the influence of particular factors on the initial stage of film growth. Its deficiency rests in that it employs thermodynamic concepts and quantities which apply to macroscopic systems. These concepts may only be employed in cases where the critical nucleus contains more than $\sim 100$ atoms. In many cases, however, it turns out that the critical nucleus has a radius of several tenths of nm and consists therefore of only a few atoms. In such cases, it is necessary to start with a different theory.

### 4.2.2 Statistical (Atomistic) Theory of Nucleation

The statistical theory describes the nucleation process when the critical nucleus consists of a very small number of atoms (from 1 to 10). It has been elaborated by Walton and Rhodin [27]. The nuclei are considered as small assemblies, account is taken of the bonds between individual particles and the substrate, and the nucleus is described in terms similar to those used for a macromolecule. These considerations, which we shall not set out in detail here, lead to the following expression for Gibbs's energy of critical nucleus formation

$$\Delta G_i^* = \Delta E_{i0}^* + i^* kT \ln\left(\frac{n_0}{n_1}\right) \qquad (4.24)$$

where $E_{i0}^*$ is the energy of disintegration of the critical nucleus containing $i^*$ atoms at the absolute zero of temperature and $n_0/n_1$ is the ratio of the number of adsorption sites to the concentration of adatoms.

The expression obtained in this case for the concentration of critical nuclei does not involve macroscopic quantitites such as $\sigma$, $\vartheta$, $\Delta G_v$:

$$\frac{n_i^*}{n_0} = \left(\frac{n_1}{n_0}\right)^{i^*} \exp\left(-\frac{E_{i0}^*}{kT}\right) \tag{4.25}$$

and the nucleation rate is given as

$$J = N\downarrow a^2 n_0 \left(\frac{N\downarrow}{\nu n_0}\right)^{i^*} \exp\left[\frac{(i^* + 1)\, Q_{\text{des}} - Q_{\text{dif}} + E_{i0}^*}{kT}\right] \tag{4.26}$$

The plot of $\ln J$ as a function of $1/T$ can be used for comparison of theory with experiments since it involves measurable quantities. Although the theory does not yield explicit expressions for the calculation of $i^*$ and $E_{i0}^*$, these quantities may nevertheless be determined by comparing the experimental results with the theoretical ones for different values of $i^*$ and $E_{i0}^*$. Fig. 45a

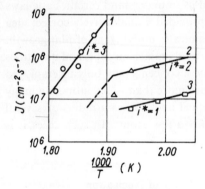

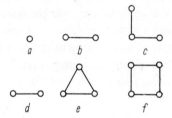

*Fig. 45a:* Experimentally observed nucleation rate of Ag on (100) plane of NaCl cleaved in vacuum (Rhodin). Deposition rate: (1) 6 . $. 10^{13}$ cm$^{-2}$ s$^{-1}$; (2) 2 . $10^{13}$cm$^{-2}$ s$^{-1}$; (3) 1 . $10^{13}$ cm$^{-2}$ s$^{-1}$.

*Fig. 45b:* Critical nuclei and the smallest stable embryos on the surface of an fcc crystal.

displays results obtained for Ag evaporated onto a single crystal of NaCl in ultra-high vacuum. At very low temperatures of substrate or very high supersaturations, a single atom already represents the critical nucleus. A stable cluster is formed by joining of another atom and the cluster then grows spontaneously. At higher temperatures a pair of atoms with a single bond per atom ceases to be a stable cluster. At this temperature the stable cluster is represented by a group of atoms with at least two bonds per atom (Fig. 45b, where a, b, c are the critical nuclei and d, e, f are the corresponding smallest stable clusters on the surface of a crystal with fcc structure). Since this theory considers the arrangement of individual atoms on the substrate surface, it provides information on the structure and orientation of growing clusters.

As the number $i^*$ of atoms in the critical nucleus depends on the temperature of the substrate, there exists a temperature at which a transition occurs from the $i^*$-atom critical nucleus to the $(i^* + 1)$-atom nucleus.

To give an example: for transition from a 2-atom to a 3-atom stable cluster, the critical temperature is given as

$$T = - \frac{Q_{\text{des}} + \frac{1}{2}E_3}{k \ln \left( \dfrac{N\downarrow}{\nu n_0} \right)} \tag{4.27}$$

($E_3$ is the energy needed for dissociation of a 3-atom nucleus into single particles). The temperature has great significance for the epitaxial growth of films; we shall discuss this in detail later.

The basic equations of the capillarity theory (4.21) and of the statistical theory (4.26) have a similar form, which is understandable since the fundamental principles used in their derivation are identical. The difference lies in the fact that the capillarity theory employs the concept of a continuously varying surface energy and introduces macroscopic thermodynamic quantities, whereas the statistical theory takes into account the discontinuous nature of changes in the binding energy with the addition of a particle to the cluster.

Somewhat different is the statistical theory developed by Zinsmeister (1965 – 69), which derives from Frenkel's theory of condensation (1924). The basic ideas are approximately as follows: To atoms impinging on a substrate, a certain mean duration and mobility may be assigned. During the staying time there occur collisions which result in the formation of pairs and later in the growth of greater aggregates. The reverse process of aggregate disintegration and re-evaporation proceeds simultaneously.

Let $N_i$ be the concentration of the aggregates containing $i$ atoms, $q'$ the number of atoms impinging on a unit area per unit time, $\alpha_i$ the coefficient characterizing re-evaporation of an $i$-atom cluster, $\beta_i$ the coefficient characterizing the disintegration of $i$-atom aggregates, $\tau_a$ the time constant characterizing the re-evaporation, $w_{ij}$ the collision factor pertinent to the collision of $i$- and $j$-atom clusters, (is dependent on the coefficient of surface diffusion) $\sigma_i$ the coefficient characterizing the addition of further particles to the $i$-atom cluster resulting from direct impingement from the gaseous phase. Then it is possible to set up a system of equations describing the film growth:

$$\frac{dN_1}{dt} = q' + \sum_{i>1} N_i \beta_i - \frac{N_1}{\tau_a} - N_1 q' \sigma_1 - N_1 \sum_{i>1} w_{1i} N_i$$

$$\frac{dN_2}{dt} = \frac{1}{2} w_{11} N_1^2 + N_1 q' \sigma_1 + \sum_{i>2} N_i \beta_i' - w_{12} N_1 N_2 - N_2 q' \sigma_2 - N_2 \alpha_2$$

$$\frac{dN_3}{dt} = w_{12} N_1 N_2 + N_2 q' \sigma_2 + \sum_{i>3} N_i \beta_i'' - w_{13} N_1 N_3 - N_3 q' \sigma_3 - N_3 w_3$$

$$(4.28)$$

The terms containing $\sigma_i$ are mostly negligible. The $w$'s may be determined from kinetic considerations (for not too high coverage). Since the most prominent role in the dissociation belongs always to the process with smallest dissociation energy, the most significant dissociation will be that of pairs.

The coefficient of pair dissociation may be roughly evaluated as follows: In a solid each atom has 12 neighbors on the average. To effect evaporation, a certain sublimation heat $Q_{subl}$ must be applied. Thus the dissociation of a pair should require $1/6$ of that amount $(E_d)$. Experiments reveal, however, that the real situation is different. The commonly used metals such as Ag, Au, Cu exhibit a much higher pair-dissociation energy $(Ag - 1.6\,eV, Au - 2.23\,eV, Cu - 1.9\,eV)$. On the other hand, some materials which are known to condense with difficulty $(Hg, Cd)$ have a lower dissociation energy than estimated. This means that in the former case the atom pairs are very stable systems whose dissociation is almost impossible at normal temperatures; the pair is thus a stable cluster. In the case of hard-condensing materials with small $E_d$ the critical nucleus must have $3-10$ atoms. However, $E_d$ may change during the adsorption on the surface of the substrate, especially when the adsorption energy is high.

Thus, if we neglect the dissociation and re-evaporation of pairs and the direct capture of particles from the gaseous phase, it is possible to replace the equation system $(4.27b)$ by a much simpler one:

$$\frac{dN_2}{dt} = \frac{w}{2} N_1^2 - w N_1 N_2$$

$$\frac{dN_3}{dt} = w N_1 N_2 - w N_1 N_3$$

$$\frac{dN_i}{dt} = w N_{i-1} N_1 - w N_1 N_i \qquad (4.29)$$

(All collision parameters are taken equal here, $w_{ij} = w$).

The system may be solved under certain simplifying assumptions. The results are shown in Fig. 46. It is readily seen that bigger aggregates reach their equilibrium concentrations only after the smaller ones. It is also

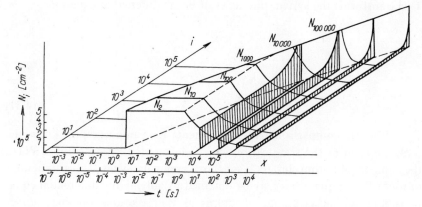

*Fig. 46:* The time variation of condensation according to Zinsmeister $(q = 10^{13}$, $\alpha = 10^{-7}$, $w = 2 \cdot 10^{-3})$. Concentration of aggregates $N_i$ as a function of real $(t)$ and synthetic $(x)$ time and number of atoms in aggregate.

possible to derive the time variation of the condensation coefficient and the time variation of the amount condensed as a function of the condensation conditions.

While normal condensation is an activated process which starts only when a certain activation energy is supplied for the formation of critical nuclei, the condensation of thin films at high degrees of supersaturation is, according to the aforementioned theory, a simple precipitation which does not need any activation energy. The fact that the condensation sometimes occurs only far below the condensation temperature is due to secondary factors, e.g. evaporation of whole aggregates (which has so far been neglected). The mobility of bigger aggregates is another factor which has been neglected. This mobility decreases with increasing size. Thus the theory should at least take account of the mobility of the twins. The results of calculations can be summed up as follows: The mobility of twins need not be considered if $w_2 < w/1000$. When $w_2 > w/1000$, inclusion of the mobility always leads to narrowing of the size distribution of aggregates and to a decrease in their density $G$.

The solution of the simplified equation system (4.29) is preceded by introduction of 'synthetic' time by the relation

$$x = w \int_0^t N_1(t)\, dt \tag{4.30}$$

$$\frac{dN_i}{dt} = \frac{dN_i}{dx} \cdot \frac{dx}{dt} = N_i' \cdot w \cdot N_1$$

This transforms the system into a set of linear differential equations

$$N_2' + N_2 = \frac{N_1}{2}$$

$$N_3' + N_3 = N_2$$

$$N_i' + N_i = N_{i-1} \tag{4.31}$$

For the comparison with experiments it is important to know the time variation of the size distribution of clusters (i.e. the quantities $N_i(t)$, the numbers of $i$-atom clusters — or those having diameter $D$ — as functions of time). These quantities may be obtained by solving the differential equations. Fig. 46 shows graphic diagrams of the solutions obtained under certain simplifying assumptions.

The theory predicts dependences of important quantities on basic condensation parameters: the density of aggregates $G = \sum_2^\infty N_i(t)$ increases with the evaporation rate $q$ $[\text{cm}^{-2} \text{ s}^{-1}]$ as $\sim q^{1/3}$. The maximum cluster diameter $D_{\max}$ decreases as $1/q^{1/9}$. Thus an increase in the evaporation rate results in the formation of a higher number of smaller aggregates.

The collision factor $w$ $[\text{cm}^2 \text{ s}^{-1}]$ determining the mobility of adsorbed atoms is related to the aggregate density and maximum cluster size by the relations: $G \sim 1/w^{1/3}$, $D_{\max} \sim w^{1/9}$, i.e. an increase in the surface mobility of particles leads to formation of a smaller number of bigger clusters.

Finally, the temperature of the substrate $T_s[\text{K}]$ influences mainly $\tau_a$ which may above all affect the character of the initial stages of condensation. It becomes less important in later condensation stages when it makes itslef felt only through the temperature dependence of the collision factor $w = w(T_s)$.

It is thus seen that the theory yields results which are in qualitative accord with observed dependences. In addition it provides greater possibilities for quantitative comparisons because it gives the time variation of the cluster size distribution. The simplifying assumptions introduced in the original form of the theory have been replaced in subsequent papers by more sophisticated assumptions which better correspond to reality.

### 4.2.3 Influence of Individual Factors on Nucleation Process

Due to the exponential dependence of $J$ on $\Delta G^*$ (or $E_i^*$), the nucleation rate depends on supersaturation and this in turn depends on the temperature. As we have seen from Fig. 44, $J$ changes very rapidly from negligible values to very high ones. Critical supersaturation corresponds to a certain critical temperature.

The theories expounded are valid only for 'steady-state' values of $J$, i.e. when the mean spacing between the nuclei is greater than the diffusion length $\overline{X}$. When the internucleus spacing just equals $\overline{X}$, the density of nuclei reaches a maximum and increases no further. The nuclei grow thereafter only by annexing additional particles by surface-diffusion. This saturation density of nuclei is independent of the impingement rate, provided the impinging atoms are accommodated instantaneously, their momentum is not significant and the impingement flow is smaller then the diffusion flow. Under these conditions, the saturation density is given by

$$N_s = n_0 \exp\left(-\frac{Q_{des} - Q_{dif}}{kT}\right) \qquad (4.32)$$

As has been shown, only those particles which remain adsorbed on the surface and condense there can take part in a film formation. Their number relative to the total amount impinging is given by the condensation coefficient. This coefficient decreases with increasing substrate temperature and decreasing binding energy of the adsorbate to the substrate. It depends also on the coverage and so it mostly increases during evaporation and approaches unity at the completion of substrate coverage. Table 9 illustrates

| The Condensation Coefficient for Cd on Cu (25 °C) | | | | *Table 9* |
|---|---|---|---|---|
| Deposit of Cd (nm) | 0.08 | 0.49 | 0.6 | 4.24 |
| s | 0.037 | 0.26 | 0.24 | 0.60 |

this fact for the condensation of cadmium on a copper substrate at a temperature of 25 °C. The differences in the condensation coefficient caused by temperature variations and the substrate properties are set out in Table 10.

The coefficients have been generally measured in the initial stages of condensation (i.e. at minimum resolvable film thickness).

The condensation coefficient depends considerably on the presence of adsorbed layers on the surface of a substrate. The coefficient has been found to be practically equal to unity for the condensation of Cd on an atomically clean tungsten surface (at pressures of $\sim 10^{-10}$ torr), whereas at residual gas pressures of $\sim 10^{-5}$ torr, it is much lower and critical super-saturation increases by several orders. In some cases, however, the adsorption of gases on the surface facilitates condensation (e.g. tin or indium condensation on glass is facilitated by adsorption of oxygen). Thus the influence of adsorbed layers depends on the particular condensate-substrate combination.

The Condensation Coefficients for Some Combinations of Materials

*Table 10*

| Condensate | Substrate | Substrate temperature (°C) | $s$ |
|:---:|:---:|:---:|:---:|
| Au | glass, Cu, Al | 25 | 0.90 — 0.99 |
| | Cu | 350 | 0.84 |
| | glass | 360 | 0.50 |
| | Al | 320 | 0.72 |
| | Al | 345 | 0.37 |
| Ag | Ag | 20 | 1.0 |
| | Au | | 0.99 |
| | Pt | | 0.86 |
| | Ni | | 0.64 |
| | glass | | 0.31 |

If $\alpha < 1$, the accommodation on the surface is insufficient, i.e. a state of thermodynamic equlibrium is not attained. In such cases, unusually high values for critical supersaturation have been observed. This fact suggests that the particles on the surface conserve a part of their energy, i.e. they remain 'hot' relative to the surface temperature so that their effective temperature (corresponding to that energy), rather than the substrate temperature, should be used for calculation of critical supersaturation. We shall return to the influence of the energy of impinging particles when dealing with formation of films by cathode sputtering.

The existence of three-dimensional disconnected islands on the substrate surface established by electron microscopy indicates the existence of a nucleation barrier (see Sect. 4.2.1) and the growth of islands primarily due to surface-diffusion. A two-dimensional monolayer arises only in exceptional cases: (a) if the nucleation barrier is small and the particle energy is high, (b) if nucleation occurs in a narrow vicinity of the impingement site on the substrate (i.e. for low substrate temperature or high-melting-point evaporation material). When the latter condition is satisfied, a high concentration of nucleation centers arises ($10^{15}$ cm$^{-2}$), so that the film is nearly continuous from the very beginning. The equilibrium density of the centers for metals deposited on insulating substrates is usually $10^{10}$ to $10^{12}$ cm$^{-2}$, which corresponds to 10 to 100 nm spacing. The sizes of the islands at a given growth stage are distributed around a certain mean value.

The density of nucleation centers may be affected also by certain external agents, e.g. by the presence of electric charge on the film which lowers the nucleation barrier and increases the binding energy, or by using so-called 'pre-nucleation'. The latter consists of the evaporation of an ultrathin layer (thinner than a monolayer) of some metal (usually one with a high melting point) onto a substrate on which a continuous film of another substance is to be deposited. The preliminarily evaporated layer facilitates the birth of nucleation centers because it increases the binding energy and it is possible by choosing a suitable combination of substances to attain a continuous coverage even with a very small thickness.

As we have noted, the nucleation occurs preferentially at the sites where the crystal lattice exhibits some irregularity, e.g. at the steps of the lattice. This is utilized in the technique of decoration.

The occurrence of preferred nucleation sites gives rise to the question of the nature of condensation centers in general. Some authors maintain that the centers are always related to a crystal defect, e.g. to the emerging points of dislocations. On the other hand, it has been established from experiments that the nuclei are randomly and isotropically distributed and that their number is no doubt considerably smaller than that of the adsorption sites but considerably higher than the density of defects on the surface of a typical single-crystal plane (at epitaxial growth). It is therefore more likely that nucleation is a random process and that it results from statistical fluctuations in the supersaturation. Another objection to the hypothesis of defect-initiated nucleation may be seen in the fact that the nucleation density is approximately the same for various substrates so that the defect density should be the same for all of them, which is improbable. Further, the logarithm of the density of nuclei is inversely proportional to the temperature whereas the density of point defects increases with temperature.

From the theoretical considerations mentioned in Sects. 4.2.1 and 4.2.2, we can derive the dependence of nucleation on the substrate temperature. Differentiation of the equations for the radius of the critical nucleus (4.11) with respect to the temperature at a constant rate of impingement and substitution of typical values results in a positive derivative, which means that the size of the critical nucleus increases with increasing temperature. Consequently, the film preserves its island character until higher values of thickness are reached. During metal-on-metal nucleation, in which there is sometimes no energy barrier at all, the barrier may arise at an elevated temperature (i.e. the film, which has originally grown in the form of a two-dimensional formation, is transformed into a film with three-dimensional island-structure).

By differentiating (4.16b) with respect to $T$ (at a constant impingement flow), we again obtain a positive value. This means that the rate of formation of super-critical nuclei decreases rapidly with the temperature. At a higher temperature, more time is needed for the growth of continuous film.

In a similar manner we may investigate the dependence of nucleation on the impingement flux (or evaporation rate).

The derivative of $r^*$ with respect to $N$ at constant temperature is negative, i.e. the rise in evaporation rate results in the diminution of the size of critical nuclei. The derivative of $G^*$ with respect to $N$ at constant temperature is also negative so that the rate of nuclei formation increases with the evaporation rate. Thus, at higher evaporation rates the continuous film may be formed at a smaller average thickness. The relationship is, however, a logarithmic one so that the evaporation rate must be changed considerably before achieving the desired effect.

As will be obvious from the theory the surface-diffusion coefficient does not affect the size of the critical nuclei but it does affect their rate of formation, which decreases exponentially with increasing activation energy of diffusion. If the energy is too high an island grows only by direct impingement of vapor particles and the aforementioned considerations are, of course, no longer valid.

For the activation energy of metal-on-metal diffusion, the following relation holds in many cases:

$$Q_{dif} \approx 0.25 \, Q_{des} \qquad (4.33)$$

Not much is known about the diffusion on nonmetal substrates. It is clear, however, that the activation energy of surface-diffusion cannot be higher than the activation energy of desorption.

The latter significantly affects both the size and formation rate of critical nuclei. The higher the energy, the smaller is the critical nucleus and the higher is the nucleation rate. The binding energies vary from several tenths of eV for van der Waal's forces to several eV for metal binding. The film-substrate binding is sometimes of a chemical nature (e.g. evaporation of Al onto glass produces an interfacial layer of oxide). This leads to a considerable increase in the binding energy and lowering of the nucleation barrier. The tendency for the formation of islands decreases and this plays a role in the deposition of Al films on glass.

### 4.2.4 Some Experiments for Verification of Nucleation Theories

We have seen earlier the capillary theory predicts the existence of a threshold nucleation frequency and critical adatom (adsorbed atom) concentration.

The equation for the nucleation frequency (4.21) may be rewritten as

$$J = \beta \exp \frac{Q_{des} - Q_{dif}}{kT} \exp \frac{-\Delta G^*}{kT} \tag{4.21}$$

where $\Delta G^*$ is a function of $\Delta G_v$ and that in turn is a function of the supersaturation (see (4.16b) and (4.9), (4.22)), and $\beta$ is a coefficient independent of the temperature.

For experimental verification of this theory two approaches are available: ($a$) the substrate temperature is kept constant and the supersaturation is varied; then

$$J = \beta' \exp\left(\frac{-\Delta G^*}{kT}\right) \quad \text{i.e. } \ln J \sim \frac{1}{T}$$

or ($b$) both quantities are varied. Then it is convenient to put (4.21) into logarithmic form and express $(\Delta G_v)^2$ as a function of $kT \ln (\beta/J)$. If the theory is valid, the dependence is linear and is called a Pound-plot. A whole number of nucleation experiments has been evaluated in this way an satisfactory results have been produced. Results for nucleation of Ag on NaCl reproduced in Fig. 45a may serve as an example.

As we have mentioned already in Sect. 4.2.2, it is possible to determine some parameters occurring in the atomistic theory by plotting $\ln J$ vs. $1/T$.

According to theoretical assumptions nucleation begins when the adatom concentration reaches a certain level. The assumption has been verified by Gretz with his experiment with a field emission microscope (see Sect. 4.2.41). He deposited various metals (Zn, Au, Cd, Ni) by means of evaporation on the tungsten tip of the microscope. At a given temperature and impingement flow $J_v$ he recorded a time $t$ needed for the formation of a three-dimensional cluster of evaporated substance as indicated by the appearance of a light spot on a fluorescent screen. He obtained a linear dependence of $J_v$ on $1/t$ (i.e., $J_v t = $ constant, which is just the critical concentration of the adsorbed particles). This confirms the validity of the classical nucleation theory.

Another technique employs a microbalance (see Sect. 3.1.1). In this way, Cinti has measured the condensation coefficient of Ag on quartz as a function of time, substrate temperature and impingement flow. Condensation has been found to be a monotonic function of time and neither the critical supersaturation nor the critical intensity of the impingement flow necessary for the initiation of the nucleation have been observed. The critical radii of nuclei derived from the experiments on the basis of the classical theory are of the order of a few tenths of nm. This suggests that the statistical theory should be applied in this case rather than the capillarity theory.

Recently a mass spectroscopy method has come into use for the investigation of the nucleation process. The surface on which a given flow is impinging is placed so that the particles emitted from it may reach directly the ionization chamber of the ion source of a sensitive mass spectrometer. In the experiments of Hudson, Nguyen Anh and Chakraverty, a monopole mass spectrometer was used. This apparatus has made it possible to detect static partial pressures of $\sim 5 \cdot 10^{-12}$ torr, evaporation rates corresponding to a vapor pressure of $\sim 10^{-8}$ torr and desorption rates of $\sim 10^{13}$ molecules/cm$^2$ s. The transient process has been observed on a sudden change of impingement flow, namely on the opening or closing of a shutter.

On the opening of a shutter the signal corresponding to the rate of desorption from the surface increases exponentially till reaching equilibrium; after closing it the signal decreases again in a roughly exponential manner. The time constant of the process naturally depends largely on the substrate temperature.

The systems investigated were Cd-on-W and In-on-Si. It has been found in the former case that at elevated substrate temperatures a strongly bound phase arises and a double monolayer may be formed without the occurrence of condensation. At lower temperatures a weakly bound phase appears, which may form even tens of monolayers without condensation. The critical supersaturation needed for initiation of the condensation is determined by exposing the substrate, which is heated to a high temperature, to a molecular beam. The substrate temperature is gradually lowered and the desorption rate is recorded. At a certain temperature, $T$, a sudden drop occurs in the rate, this corresponds to the transition from the equilibrium desorption to the equilibrium corresponding to the evaporation of the condensed phase. According to the results of Hudson, the critical supersaturation is roughly a linear function of $1/T$ and the logarithm of the coverage at the onset of nucleation is a decreasing linear function of $1/T$.

For the system In on Si, the mean life-times of the adsorbed particles have been measured in a similar manner at various temperatures. At lower temperatures ($< 1000$ K), the curves of the transient process are not exponential and the time constants calculated from the decrease and increase starting points of the curves differ from each other: they correspond to the time of equilibrium coverage and that of very small coverage. From these data the dependence of the concentration of adsorbed atoms on $1/T$ has been calculated together with the dependence on surface concentration of mean life-time of In atoms on an Si surface (always at a constant impingement flow). It has been found at the same time that both the condensation and accommodation coefficients are equal to unity. It has been also found that condensation occurs when the pressure equivalent to the impinging flow is lower than that of In vapor at the given temperature (i.e. it occurs at an undersaturation of $\sim 0.9$).

Some of the results of the experiments are rather surprising and cannot be explained on the basis of existing theories, particularly the high coverage in the Cd-on-W system preceding the nucleation proper and the relatively low supersaturation (about $1.6 - 1.8$ instead of $10^6 - 10^{19}$). Nor is it possible to apply the atomistic theory.

In the In- on Si-system the nucleation occurs even at undersaturation. Here, the surface states on silicon may possibly play some role by representing nucleation centers with a high binding energy.

These experiments thus demonstrate that there are cases of nucleation which lie outside the scope of current theories and which thus may become a stimulus for further theoretical investigations.

## 4.3 Growth and Coalescence of Islands

For the final structure of a film further stages of growth are important, namely, the growth of individual islands and especially their coalescence.

Growth occurs in the main by surface diffusion of adsorbed atoms (adatoms) and their annexation to the surface of the already existing nuclei. The process is sometimes described in forms of a two-dimensional gas of particles adsorbed on the surface: an island of a radius $r$ may be in equilibrium only with a concentration of adatoms given by the Gibbs-Thomson equation:

$$n_t = n_{eq} \cdot \exp \frac{2\sigma V_m}{kTr} \qquad (4.34)$$

where $V_m$ is the volume of an adatom, $n_{eq}$ the concentration of adatoms corresponding to the equilibrium pressure of the vapor of island material at the temperature $T$ determined by a relation analogous to (4.7),

$$\frac{p_{eq}}{\sqrt{(2\pi m k T)}} = n_{eq} v \exp\left(\frac{-Q_{des}}{kT}\right) \tag{4.35}$$

$\sigma$ is the interfacial energy per unit area.

If the mean concentration of adatoms $n$ is greater than the equilibrium concentration $n_t$, the island will grow; if it is smaller, the island will disintegrate.

According to Chakraverty, the growth rate of an island is always limited by the slower of the two processes cooperating in the growth, i.e. the surface diffusion and interface transfer.

In the case of the process limited by surface diffusion, the theory proceeds from the solution of Fick's second law on the assumption that the concentration of adatoms varies in the island neighborhood from $n'$ at the island fringe, i.e. at the distance $R = r \sin \vartheta$ from the center of the island, to $\bar{n}$ at the distance $R = l \cdot r \cdot \sin \vartheta$, where $l$ is the screening distance, i.e. the distance at which the concentration again reaches its mean value.

For the growth rate, the relation obtained is

$$J_s = \frac{2\pi D_s}{\ln l}(\bar{n} - n') \tag{4.36}$$

where $D_s$ is the coefficient of surface diffusion.

The rate of interface transfer is determined by the interface area and the difference in the numbers of atoms which join and leave the island. Thus

$$J_t = 4\pi r^2 \phi_1(\vartheta) \beta_0 (n' - n_t) \tag{4.37}$$

where $\beta_0$ is a probability coefficient dependent on the temperature. At equilibrium, both rates must be equal,

$$J_s = J_t = J \tag{4.38}$$

so it is possible to eliminate the unknown quantity $n'$ from (4.36) and (4.37) and thus obtain the following expression for $J$:

$$J = \frac{(2\pi D_s/\ln l)\,\beta_0\,4\pi r^2\,\phi_1(\vartheta)}{(2\pi D_s/\ln l) + \beta_0\,4\pi r^2\,\phi_1(\vartheta)}(\bar{n} - n_t) \tag{4.39}$$

The time variation of the island volume is

$$\frac{d}{dt}\left[\frac{4}{3}\pi r^3\,\phi_2(\vartheta)\right] = J V_m \tag{4.40}$$

from which the time variation of the radius is obtained in the form

$$\frac{dr}{dt} = \frac{J V_m}{4\pi r^2\,\phi_2(\vartheta)} \tag{4.41}$$

$(\phi_1(\vartheta))$ and $\phi_2(\vartheta)$ are geometrical factors which occur in the capillarity theory of nucleation).

Using these concepts. Chakraverty derived an expression for the time distribution of islands at a given time. Similar results may be also obtained by using the statistical theory of Zinsmeister (see Sect. 4.2.2).

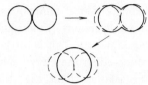

*Fig. 47:* Schematic illustration of coalescence of two nuclei.

During the following stage of film growth some islands come into mutual contact and coalescence ensues as illustrated in Fig. 47. The islands behave during coalescence like two droplets. By this process the large interfacial energy possessed by the system when comprised of isolated islands is decreased. The nuclei growing on a substrate may have various crystallographic orientations and, in general, various conditions for growth. Large islands grow faster and small ones partly disappear due to the coalescence with larger ones. At each instant there is a certain size distribution of the islands. It has been found that the distribution is not a Guassian one. It may be observed best by observing film growth directly in the viewing field of an electron microscope, as has been carried out by e.g. Pócza and Barna. An interesting phenomenon is that in some cases the islands assume a crystallite shape (with pronounced crystal planes) but behave as a liquid during the coalsecence proper, after which a new crystallite is formed. The phenomenon of the liquid-crystal and the inverse phase transition may be observed by means of electron diffraction. It is believed that a certain energy is liberated by coalescence which is sufficient to effect a temporary melting of the crystallites in contact. After coalescence the temperature drops and the newborn island again crystallizes. It has been established that when two islands which are of different sizes and crystallographic orientations coalesce, the resultant crystallite assumes the orientation of the larger one.

This may substantially modify the resulting orientation of the film. It may happen that the nuclei orientations are not evenly represented so that, for example, the [100] orientation prevails over the [111] orientation. If, however, the conditions for growth are better for [111] islands, they will have a greater size when coalescing with those of the [100] orientation. The resulting islands will thus be oriented in the [111] direction so that the [111] orientation may become dominant even when [100] orientation has prevailed during nucleation proper.

Let us return again in more detail to the problem of condensation temperature and to the related problem of the existence of condensate in the liquid or crystalline form. It is necessary to find out whether the condensation occurs by the direct vapor-to-solid transformation or via the vapor-liquid-solid transformation.

Observations obtained by means of electron diffraction seem to indicate that at temperatures higher than $2T_m/3$ ($T_m$ is the melting point of the bulk material) the islands yield diffuse diffraction rings, betraying their liquid nature. This is further sustained by the spherical shapes of the islands observed in the electron microscope. At a temperature lower than $2T_m/3$, the islands yield the diffraction pattern corresponding to a crystalline structure. And, finally, at a temperature lower than $T_m/3$, the diffraction pattern is again diffuse. It would be difficult, however, to assume that a liquid state occurs at such a low temperature. The material should rather be characterized as amorphous.

These observations are explained as follows: The temperature $2T_m/3$ may be very close to the melting point of a small crystallite, which is always lower than that of the bulk material. The melting point $T_r$ of a spherical crystallite of radius $r$ is lowered in consequence of the increase in vapor pressure over the curved surface. According to the Thomson-Frenkel theory the melting point is

$$T_r = T_m \exp\left(\frac{-2\sigma V}{Lr}\right) \qquad (4.42)$$

where $\sigma$ is the liquid-solid interfacial energy, $L$ is the latent heat of fusion and $V$ is the molecular volume of the solid phase. For small differences $T_m - T_r = \Delta T$ one obtains

$$\Delta T = \frac{2\sigma}{r}\frac{T_m}{L}V \qquad (4.43)$$

If the radius of the critical nucleus is substituted for $r$, then

$$\frac{T_m}{T_r} = 1 + \frac{\sigma}{\sigma_{cv}}\frac{kT_m}{L}\ln p/p_r \qquad (4.44)$$

Since for many metals $kT_m/L \approx 1$ and $\sigma/\sigma_{vc} \approx 0.1$, it is necessary that $p/p_r \approx 10^2$ if the experimentally found value $T_m/T_r \sim 1.5$ is to be obtained. The value is, of course, many orders lower than the observed levels of supersaturation. Thus the theory does not lead to a quantitative agreement with experiment, but it provides, however, qualitative explanation of the fact that a substance may condense as a liquid even at a temperature lower than the melting point of the bulk material.

As for the threshold $T_m/3$, the value is not yet substantiated by theory. It is, however, probable that at certain lower temperatures the surface mobility of condensing particles is so low that crystallization cannot occur and the substance assumes the structure of an undercooled liquid.

## 4.4 Influence of Various Factors on Final Structure of Film

The majority of the physical properties of thin films utilized in practical applications depend to a considerable degree on the structure of the film. It is therefore important to know how particular factors may influence the structure during film growth.

As we have already seen it is not only the nucleation process which is important for the formation of the final film structure but also the process of further film growth. In addition, recrystallization, proceeding especially at elevated temperatures, may also play a role. For the formation of an oriented epitaxial film, it is not sufficient that a large number of nuclei with the given orientation arise at the onset; it is necessary that just these nuclei have optimal conditions for growth so they grow faster than nuclei of other kinds and are of prevailing significance in the recrystallization occurring during coalescence. All these stages may be influenced substantially by impurities on the substrate surface, which explains the great variety of results obtained from the actual observation of the growth of films.

The impurities are bound to the surface with various binding energies. Energies of the order of $0.1-0.5$ eV correspond to physical adsorption, and energies of $1-10$ eV to chemisorption. The influence of impurities depends also on the energy of the impinging particles. The energies for normally evaporated particles are $\lesssim 1$ eV, and for particles in cathode sputtering or those evaporated by special methods (see Sect. 2.3.25) they may be one to two orders higher. We shall be concerned in the next section with the peculiarities of the condensation of a film prepared by these methods.

Surface impurities influence not only the binding energy between the deposited substance and the substrate, and so the size and growth conditions of critical nuclei, but they may even effect a secondary nucleation while no additional nuclei are formed on the clean substrate. Further contamination may lead to the formation of nuclei on the surface of existing islands.

Impurities from the residual gas are also built directly into the film and may influence substantially its resistivity, magnetic properties, etc. At the same time, it is not so much the total pressure of the residual gas which determines this influence as the partial pressure of some of its components (e.g. oxygen, water).

Further, impurities on the substrate may considerably alter the adhesion of the film to the substrate. The adhesion is strongest when a layer of compound, e.g. oxide, may form in between the film and substrate, as happens, e.g. in the case of iron or aluminium on glass. Adhesion is much weaker if the binding consists of only van der Waals's forces. Even very thin layers of a substance adsorbed on the surface may prevent the formation of the oxide and thus modify substantially the magnitude of van der Waals's forces. Adsorbed layers of organic substances originating from the vacuum seals or the pumping fluid are especially detrimental to adhesion. Heating the substrate before deposition or a raised temperature of the substrate during deposition often improve film adhesion.

We may also note here that if no compound is formed between the film and substrate, good adhesion may be achieved by a transition layer grown by mutual diffusion. This has been demonstrated, for example, on Cd—Fe, a combination which normally shows weak adhesion. If, however, the film is prepared by cathode sputtering, the impinging particles have a higher energy so that the surface is cleared of oxide, precluding mutual diffusion, the deposition centers have higher binding energy and, finally, the particles themselves penetrate the surface to a greater depth. All these factors result in enhanced adhesion.

Besides the cleanness and structure of the surface and the residual gas pressure, the most important factors are the evaporation rate and substrate temperature. We have dealt with their effect on the nucleation process in Sect. 4.2.4. In addition, the evaporation rate considerably influences the content of the impurities in the film itself as follows from equations (2.4) and (2.5).

### 4.4.1 Special Properties of Films Deposited by Cathode Sputtering

We have mentioned earlier the fact that particles impinging on a surface in cathode sputtering have substantially higher energies than particles produced by evaporation (tens, sometimes even hundreds, of eV). Such particles obviously behave differently on the surface than slower ones. Above all, they usually retain a considerable part of their energy so that they are able to move overt the surface even at temperatures at which evaporated particles would in practice be localized. On the other hand, those particles with the highest energy may create defects at the site of impingement that will therefore have a higher binding energy than adjacent areas of the substrate and thus will become sites of preferential nucleation.

Independently of theoretical considerations, it was found earlier that films (e.g. Ag) prepared by cathode sputtering coalesce into continuous film at smaller mean thicknesses than similar films prepared by evaporation.

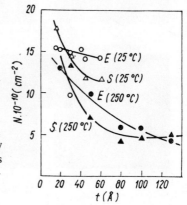

*Fig. 48:* The variation of the island density N with thickness of Ag film on mica. The films have been prepared by: E — evaporation; S — cathode sputtering (Chopra).

Further research, especially that of Chopra, confirmed the results and demonstrated at the same time that it is possible by cathode sputtering to prepare epitaxial films of a high quality similar to those obtainable by evaporation *in vacuo*. Moreover, owing to the greater surface mobility of cathode-sputtered particles, the films have properties similar to those of films evaporated onto the substrate at higher temperatures. The higher nucleation density due to the production of point defects on the substrate surface is enhanced by the influence of the electric charge carried by the particles. The charge also enhances the inter-island diffusion and hence accelerates coalescence. Results of the measurement of island density N vs. film thickness for Ag prepared on mica by the evaporation method $(E)$ and cathode sputtering $(S)$ is illustrated in Fig. 48.

As illustrated, the nucleation density is higher in S at the beginning, which is due to point defects and the influence of the electrostatic charge (the nucleation density is of the same order of magnitude for both S and E, i.e. $10^{11}$ cm$^2$). For E at a temperature of 25 °C the density remains practically constant owing to the low mobility, but at 250 °C it decreases. For S the density decreases owing to high mobility even at room temperature and at 250 °C it very soon reaches a constant value, which means that film S is continuous already at a smaller thickness.

Epitaxial growth of evaporated films occur (see Sect. 4.6) only above a certain epitaxial temperature. This depends on both substrate and evapo-

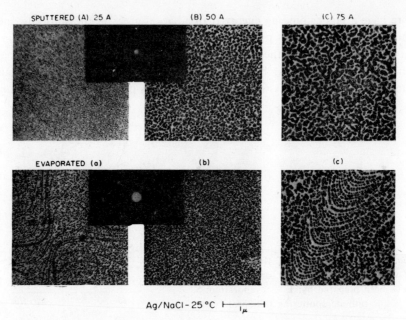

Fig. 49: Films of Ag on NaCl prepared by evaporation and Ar-sputtering at substrate temperature of 25 °C (With permission of Dr. Chopra).

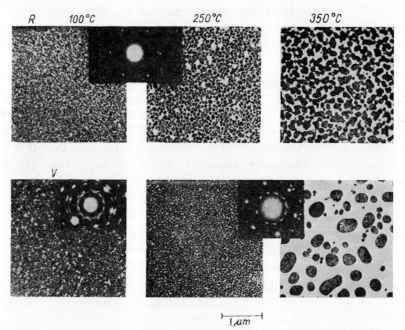

Fig. 50: 10-nm-thick films of Ag: V — evaporated; R — Ar-sputtered on mica at various substrate temperatures (With permission of Dr. Chopra).

rated material and also on the evaporation rate. It has been established that cathode-sputtered films show epitaxial growth at far lower temperatures, sometimes at temperatures below zero. The first row of Fig. 49 shows three electron-microscope illustrations of Ag films deposited by cathode sputtering in Ar onto a single crystal of NaCl at 25 °C; below them are films of the same thickness but prepared by evaporation. In both cases the electron diffractograms are also shown. A certain decorative effect may be observed on the thinnest evaporated film which is not present on the sputtered one owing to higher energies of the particles. Further, there is a difference in the coverage of the thickest films, the sputtered sample showing a greater tendency to uniform coverage of the surface.

In the further illustrations the difference between evaporated and sputtered films can be seen still more clearly. In the first row of Fig. 50 are shown films 10 nm thick, sputtered onto the substrate at various temperatures. Below them are shown analogous illustrations of the evaporated ones. Now, it can readily be seen that at higher temperatures evaporated film forms large droplets on the surface and does not manifest the tendency for growth of continuous film, in constrast to sputtered film. The electron diffractograms (insets) show at the same time that the cathode-sputtered films achieve a perfect monocrystalline structure even at the lowest temperature, whereas in evaporated films partial orientation occurs only after a higher temperature is reached.

*Fig. 51:* Ag films of various thickness deposited on mica by He cathode sputtering at 25 °C (With permission of Dr. Chopra).

The question of whether or not these differences in the structure are actually due to the impingement energy of the particles has been solved by sputtering in helium instead of sputtering in argon. Helium has much lighter atoms which carry substantially lower energy to the substrate than argon atoms, provided other conditions are the same. The result of the experiment is shown in Fig. 51. Films have been sputtered at the substrate temperature of 25 °C, so the conditions are similar to those of the first row of Fig. 49. The electron diffractogram indicates, however, that the film is now only partially oriented, in contrast with that sputtered in argon. This is proof that the energies of the impinging particles play an important role in the determination of film orientation.

Similar results could be expected in films prepared by evaporation by a laser beam or by the exploding-wire method.

## 4.5 Crystallographic Structure of Thin Films

We have already come across the fact that films may have various crystallographic structures depending on the actual conditions under which they grow. There are essentially three different groups of films: amorphous, polycrystalline and monocrystalline.

Amorphous films are usually formed by such elements as C, Si, Ge, Se, Te, some compounds of Se and Te and by some oxides, if they are prepared with the substrate at room temperature. These are generally the systems characterized by a low surface mobility of adsorbed particles so that the disordered state is frozen before the particles are able to reach the most preferable energetic sites corresponding to the crystallographic structure of the given substance. The amorphous state is therefore a metastable state and such films easily recrystallize with accompanying liberation of energy.

The reduction of the surface mobility of particles which leads to the formation of amorphous films may also be achieved in systems which under normal conditions form crystalline films. The reduction is effected by, for example, admixtures to the residual gas. Oxygen at a pressure of $10^{-4}$ to $10^{-5}$ torr prevents the formation of crystallites of easily oxidized substances because the resulting oxide prohibits the coalescence of islands and hence the formation of larger crystallites.

In other films the stabilization of the amorphous state is achieved by different impurities. For example, one atomic percent of nitrogen in the working gas is sufficient in cathode sputtering of W, Mo, Ta and Zr.

There is, in general, an increased tendency for the formation of an amorphous film whenever fast cooling of the condensate is adopted. Pure metals, however, in contrast to elements forming covalent bonds, form crystallites even at liquid helium temperature, though very small ones ~5 nm). The diffraction pattern of such films corresponds to a polycrystalline structure; it is, of course, a highly disordered state and this fact is shown by the high residual resistance of the film (see Sect. 6.2.1).

The cause of the difference beween metals and substances with covalent bonding is that the small coordination number (the number of nearest neighbors) in covalent-bond materials (4) requires a relatively large displacement of particles if they are to occupy positions which correspond to the crystal structure. Metals, on the contrary, crystallize in close-packed structures in which the arrangement of the particles does not differ as much from the random arrangement. Besides that, they have also lower desorption energy and hence (as adsorbed particles) higher surface mobility.

On metal films it is possible to reduce mobility and obtain an amorphous structure by co-evaporation of two suitably chosen materials, e.g. silver with 16% SiO or tin with 10% Cu. Similar results may be achieved also by cathode sputtering. Such systems usually recrystallize at temperatures of $0.30 - 0.35 \, T_s$, where $T_s$ is the mean melting point of both components.

By co-evaporation of two components, it is possible to produce solid solutions of materials which do not dissolve in each other; the solutions thus produced are, of course, metastable. At a temperature of $\sim 0.3 \, T_s$ the solutions recrystallize and form a single-phase metastable alloy; at a temperature of $0.5 \, T_s$ the metastable system disintegrates and a two-phase system is established.

By this means relatively stable systems may be obtained which do not exist otherwise — not only in the form of anomalous alloys and solid solutions but also as pure metals in crystalline forms other than those usually assumed by them. For example, tungsten, molybdenum and tantalum, which crystallize in a cubic, body-centered structure (denoted as bcc) may be prepared in the thin-film form as cubic, face-centered (fcc) materials; the compounds Cds, CdSe, ZnS, ZnSe, etc., which crystallize in the wurtzite structure transform into a sphalerite one, etc. (we shall return to the crystallographic systems later). These anomalous structures are usually rather stable but may be transformed in the normal ones by elevation of the temperature, irradiation with electrons or by treatment with an electric or magnetic field. The transition from the amorphous to the normal crystalline state sometimes proceeds via a succession of metastable states.

Films with a polycrystalline structure may exhibit various crystallite sizes. If the crystallites are smaller than 2 nm, it is not possible (by means of electron diffraction) to distinguish such films from amorphous ones. Indeed, no sharp boundary exists between amorphous and microcrystalline films.

If we are to discuss the crystallographic structure of films in more detail, we should recall some basic concepts of crystallography: We speak about a crystal when the spatial arrangement of atoms or ions which constitute the substance exhibits certain symmetry and if we can build up the crystal by a periodic spatial repetition of some basic module (primitive cell). According to the symmetry of the cell, determined by three vectors $a$, $b$, $c$ of certain lengths which form the angles $\alpha$, $\beta$, $\gamma$, we divide the lattices into fundamental crystallographic systems. These are the cubic, hexagonal, trigonal, tetragonal, orthorhombic, monolinic and triclinic systems. In Table 11, the characteristic properties of the basic cells pertinent to the systems are given.

The Parameters of the Primitive Cells in Individual Systems     *Table 11*

| System | Characteristic properties of primitive cells |
| --- | --- |
| cubic | $a = b = c$; $\alpha = \beta = \gamma = 90°$ |
| hexagonal | 3 coplanar axes at 120° to each other; fourth axis $c \neq a$; $c \perp a$ |
| trigonal | $a = b = c$; $\alpha = \beta = \gamma \neq 90°$ |
| tetragonal | $a = b \neq c$; $\alpha = \beta = \gamma = 90°$ |
| orthorhombic | $a \neq b \neq c$; $\alpha = \beta = \gamma = 90°$ |
| monoclinic | $a \neq b \neq c$; $\alpha = \beta = 90° \neq \gamma$ |
| triclinic | $a \neq b \neq c$; $\alpha \neq \beta \neq \gamma \neq 90°$ |

It is sometimes convenient to introduce an elementary cell in which the atoms are not confined only to the corners. The symmetry of such a cell corresponds more closely with crystal symmetry and its edges form a simpler coordinate system than the edges of the primitive cell. Fig. 52a shows the elementary cell of the nickel lattice, drawn with a dashed line; the full line is used for the primitive cell. The elementary cell has atoms not only at the

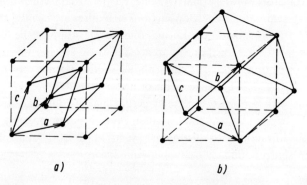

*a)*                              *b)*

*Fig. 52:* Primitive cell (solid line) and elementary cell (dashed line): (a) face-centered cubic lattice; (b) body-centered cubic lattice.

corners but also in the centres of the cube faces; we speak of the face-centered cubic lattice. In Fig. 52b, both cells of the body-centered cubic lattice are depicted in detail.

Analogously we could construct an elementary cell for the case in which the atoms are in the centers of both bases (base-centered lattice). If we consider the compound cells in all seven crystallographic systems, we obtain fourteen different lattices, called Bravais lattices (their number is not 7 × 4 because some of them are identical). The lattices are illustrated in Fig. 53.

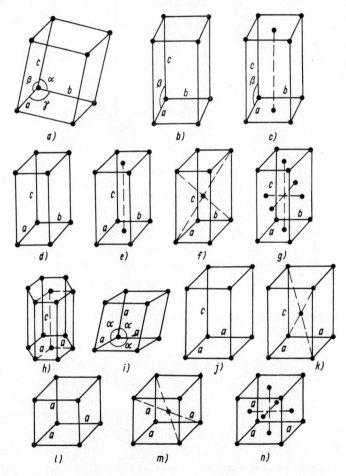

*Fig. 53:* Bravais lattices: (a) triclinic; (b) simple monoclinic; (c) base-centered monoclinic; (d) simple orthorhombic; (e) base-centered orthorhombic; (f) body-centered orthorhombic; (g) face-centered orthorhombic; (h) hexagonal; (i) rhombic; (j) tetragonal; (k) body-centered tetragonal; (l) simple cubic; (m) body-centered cubic; (n) face-centered cubic.

Crystals are restrained by crystalline planes, which are denoted by the Miller indices introduced for this purpose. Their meaning will first be explained in two-dimensional terms (Fig. 54). Let us pick up a plane which

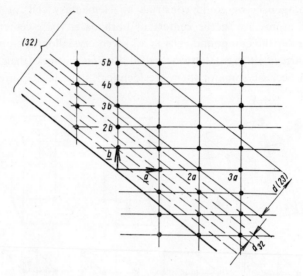

*Fig. 54:* Illustration for explanation of Miller indices.

passes through the points 2a and 3b (in general $P_1a$, $P_2b$). In an analogous manner we could specify a plane in space by intercepts $P_1a$, $P_2b$, $P_3c$, where $P_1$, $P_2$, $P_3$ are relatively prime integers (nonzero) so that the intercepts on the axes represented by them are the smallest integral multiples of the vectors $a$, $b$, $c$ for the set of planes parallel to the one selected. The Miller indices are obtained by multiplying the inverse values $1/P_1$ by the smallest common multiple of the $P_i$.s, i.e. by $P_1 P_2 P_3$:

$$(hkl) = \left(\frac{1}{P_1}, \frac{1}{P_2}, \frac{1}{P_3}\right) P_1 P_2 P_3 = (P_2 P_3, P_3 P_1, P_1 P_2) \quad (4.45)$$

If, however, the selected plane is parallel to some axis, then the pertinent $P_i = \infty$. The corresponding Miller index equals zero and multiplication by it is omitted. The indices for the example illustrated in Fig. 54 are $(hk) = (32)$.

The indices $hkl$ denote all planes parallel to the basic one which are occupied by atoms arranged in the same manner as in the basic plane. The intercepts cut by these adjacent planes are $a/h$, $b/k$, $c/l$ and the perpendicular spacing of the planes is given by

$$d_{hkl} = \frac{d(P_1, P_2, P_3)}{P_1 P_2 P_3} \quad (4.46)$$

If the vectors, *a*, *b*, *c* are perpendicular to each other (which occurs in the cubic, trigonal and orthorhombic systems), the realtion may be written as

$$d_{hkl} = \frac{1}{\sqrt{\left[\left(\dfrac{h}{a}\right)^2 + \left(\dfrac{k}{b}\right)^2 + \left(\dfrac{l}{c}\right)^2\right]}} \qquad (4.47)$$

In Fig. 55 three significant planes of the cubic system are shown together with their corresponding Miller indices.

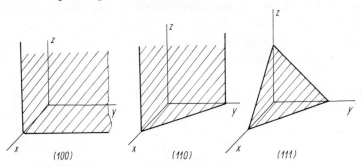

*Fig. 55:* The most important planes in a cubic system.

If we want to denote a plane intercepting some axis in the negative region, we use a bar placed above the relevant index, as for example, $(0\bar{1}0)$. If the direction of the normal to the plane is to be expressed, Miller indices are placed between brackets, e.g. $[100]$. A set of analogous planes (differing in the sign of the intercepts) is denoted by the indices of the plane with positive intercepts placed in between braces; thus $\{111\}$ denotes the set of planes $(\bar{1}11)$, $(1\bar{1}1)$, $(\bar{1}\bar{1}1)$, $(11\bar{1})$, $(\bar{1}1\bar{1})$, $(1\bar{1}\bar{1})$, $(\bar{1}\bar{1}\bar{1})$.

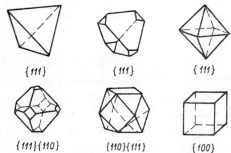

{111}    {111}    {111}

{111}{110}    {110}{111}    {100}

*Fig. 56:* Some of the most common forms of crystals.

Examples of the shapes of crystals frequently occurring in the domain of thin films, with the respective symbols the planes, are given in Fig. 56. The basic properties of crystals in thin film form are mostly the same as

108

those of the bulk material. Some special features do, however, exist. If crystallites are very small, a change $\delta d$ of the lattice constant is observed. The theory gives the following expression for the change:

$$\frac{\delta d}{d} = -\frac{4}{3}\frac{\sigma}{ED} \qquad (4.48)$$

where $\sigma$ is the surface energy, $E$ is the modulus of elasticity and $D$ the crystallite radius. The expression $4\sigma/D$ denotes the internal pressure in the crystallite, which may be either positive or negative according to the sign and may thus cause an expansion or contraction of the lattice constant. Experimental verification is difficult because it must be carried out on crystals which are entirely stressless and the technique used has to be very accurate. (The moiré method is commonly applied, see Sect. 5.2.) A contraction of the lattice constant with diminishing size of the crystallites has been established in LiF, MgO and Sn.

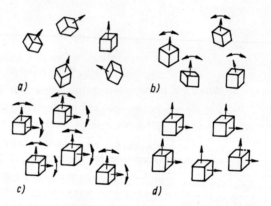

*Fig. 57:* Orientation of crystallites on substrates (double arrows indicate departures from prevailing orientation): (a) chaotic ordering; (b) single fiber texture; (c) double texture; (d) monocrystalline orientation.

In some cases when the substrate has a lattice constant approaching that of the film (the difference between them being 0.2%), 'pseudomorphism' occurs and the film assumes the structure of the substrate until a thickness of the order of 10 nm is reached. If the difference is greater (4%) and the binding between the substrate and film is strong, then pseudomorphism occurs for only the earliest atomic layers. Still greater differences are accompanied by the formation of lattice defects and dislocations.

Crystals growing on a substrate may be variously orientated. The possible orientations are shown in Fig. 57. Case (a) illustrates a completely disordered film, in which the directions of axes of individual crystallites are distributed randomly. Case (b) describes a state in which one particular axis of all the crystallites is oriented in approximately the same direction. We speak about a first-stage orientation or a single texture. The common axis is called an axis of texture or a fiber axis. Case (c) depicts so-called second-stage orientation or double texture. Finally, case (d) depicts a mono-crystalline orientation, which is very important because it includes also the case of epitaxial films (see Sect. 4.6).

The single texture, which is also called fiber texture, occurs mainly in metals and semiconductors with homeopolar bond. It may arise during nucleation, after the onset of further crystallite growth or during the complementary tempering.

There are many possible cases of texture growth depending on concrete conditions and certain empirical rules for texture formation have been derived. Phenomenological models proposed hitherto do not, however, take account of all the factors which actually influence texture growth.

Let us present two examples here: Metals with a fcc lattice have, when evaporated onto an amorphous substrate, initially the [111] texture, which transforms into the [110] at greater thickness. The metals crystallizing in the hexagonal system and having a low melting point, together with compounds of the wurtzite structure, assume a well defined [001] orientation in the nucleation stage. For perpendicular incidence the fiber axis is usually close to the normal to the substrate.

*Fig. 58:* The origin of dislocations at boundaries of crystallites.

We should mention here defects of crystal lattices in thin films, of which there are a great variety of, e.g. dislocations, point defects, crystal twins, grain boundaries, and stacking faults.

Electron-optical examinations of fcc metals by the moiré technique have revealed that there are almost no defects in separate nuclei and that they arise only after the coalescence process begins. The most frequent defects are dislocations, which arise, for example, at the boundary of two crystal regions that are somewhat angularly displaced with respect to each other (Fig. 58). The density of dislocation lines is about $10^{10} - 10^{11}$ lines/cm². These dislocations arise mainly from coalescence of large islands. If the islands are very small, they can move or rotate somewhat to eliminate the difference in orientation. If they are larger, this is no longer possible and thus dislocations arise in the region of contact. Their number increases especially at the concluding stages of the coalescence of big islands.

Dislocations are also formed on the boundary between film and substrate owing to the difference in their crystal lattices. An overgrowth of substrate defects may also occur in the film. The number of defects in the film is, however, greater by several orders than that of the substrate, so this mechanism is clearly not the main source of the defect. (In single crystals of NaCl, which often serve as substrates for monocrystalline metallic thin films, the defect density is about $10^5/cm^2$.)

The density of some types of defects usually reaches a maximum at 50% coverage; then it decreases with additional deposition and some defects may disappear in continuous films (e.g. twins or stacking faults). The number of defects may be influenced by an additional annealing of the finished film.

The fundamental 'defect' of a crystal lattice is, of course, the surface itself since it represents a cutoff of the periodicity. The surface structure in thermal equilibrium cannot in general conform to that of the bulk material.

The surface particles are arranged in two-dimensional superstructures which may be observed by methods that reveal the structure of only a few atomic layers, e.g. low-energy electron diffraction (see Chapter 5). By this method, the crystallographic symmetry of the surface structure may be determined but it is not yet possible to carry out a complete structural analysis, since the requisite theoretical foundations are still lacking.

A whole number of possible surface arrangements (ordered and disordered structures) exists for a given material, depending on the temperature. At higher temperatures the superstructures of higher orders may appear, i.e. the arrangements with a long-distance periodicity (to hundreds of nm).

## 4.6 Epitaxial Films

Epitaxy or the oriented growth of films on monocrystalline substrates is a very interesting phenomenon from a theoretical point of view and very important from a practical one.

The term denotes formation of monocrystalline films usually on monocrystalline substrates either of the same substance (e.g. films of Ge on a monocrystal of Ge), when we speak of autoepitaxy (homoepitaxy), or of other substances (e.g. Ag on NaCl), which we term hetero-epitaxy. There are also special cases of growth over an amorphous material or a liquid surface which are called rheotaxy.

The substrate has a very significant influence on the particular orientation of the growing film. Epitaxy can occur between materials of different crystal structure and of different chemical bondings, so the causes of epitaxy are apparently not simple. For example, gold can grow epitaxially on alkali halides, which have a cubic structure, on monoclinic mica and on various crystal faces of single crystals of Ge. The resulting orientation of the film, however, depends in each case on the crystal structure and orientation of the substrate. Thus fcc metals grow on NaCl in parallel orientation with [100], [110], [111] directions, but on mica they grow so that their [111] direction is parallel to the [001] of mica.

Between the planes of substrate and film, which are in mutual contact, some symmetry relations exist (more frequently of rotation than translation symmetry) which may be very complicated.

It was formerly believed that the condition for epitaxy was a degree of correspondence between the lattice constants of the substrate and those of the film. But it has been found that a small difference in the lattice constants ('misfit') is neither a necessary nor a sufficient condition for epitaxy. Epitaxy has been observed even in cases of large misfits of both positive and negative sign.

Studies of epitaxy phenomena deal mostly with the following types: metal on metal, metal on alkali halide or mica, alkali halide on alkali halide, semiconductor on semiconductor.

The fundamental parameter of epitaxial growth is the substrate temperature. For every pair of materials (provided all other conditions are constant), a certain critical epitaxial temperature exists above which epitaxy is perfect and below which it is imperfect. In addition, the epitaxial temperature for a given pair of materials depends on the evaporation rate.

It is obvious from the concepts of thin-film formation set out in this work that an increase in temperature will stimulate epitaxy: desorption of impurities will be facilitated, supersaturation lowered, and surface atoms will have more energy to attain equilibrium sites. At higher temperatures, more of the requisite activation energy (necessary for their mobility) will be supplied to adatoms and thus the recrystallization on coalescence will be facilitated as both surface and bulk diffusion will be enhanced. Higher temperatures may also contribute to the ionization of adatoms. The energy needed for epitaxial growth may also be supplied by other means. It has been observed, for example, that epitaxy may be stimulated by ultrasonic agitation. In Sect. 4.4.1 we have pointed out that the growth of a monocrystalline film is also facilitated by adequately increasing the energy of the impinging particles.

It is difficult to provide precise values of epitaxial temperatures because the temperature depends on many other parameters. Nevertheless many of these values appear to be reproducible. To give an example, the values of epitaxial temperatures for silver on several different alkali halides are shown in Table 12. It is evident that there is a certain systematic dependence on substrate properties.

Epitaxial Temperature for Ag on Alkali Halides                    *Table 12*

|            | Ag/LiF | Ag/NaCl | Ag/KCl | Ag/KI |
|------------|--------|---------|--------|-------|
| $T_e$ (°C) | 340    | 150     | 130    | 80    |

A further significant factor for epitaxy is the deposition rate R. The following equation has been found empirically

$$R \leqq A\exp\left(-Q_{\text{dif}}/kT_e\right) \tag{4.49}$$

where $A$ is a constant and $Q_{\text{dif}}$ is the activation energy of surface diffusion. This may be expressed in the following way: an adatom must have enough time to jump into a position of equlibrium before it collides with another atom.

A similar relation is derived from the Walton - Rhodin theory of nucleation for transfer from the nuclei with a single bond to those with a double bond (see Sect. 4.2.2).

In some cases epitaxy occurs only at low deposition rates whereas at higher rates crystal twins are formed (i.e. pairs of crystals planes set so that each is a mirror image of the other one). Generally the amorphous, polycrystalline and monocrystalline phases exist simultaneously in a very wide temperature range.

Another factor influencing epitaxial growth is the contamination which may be caused by an adsorption of residual gas, especially at the higher working pressures. Depending on how the adsorbed substance affects the mobility of adatoms over the substrate, the epitaxial temperature may rise or fall. Some gases (e.g. oxygen and nitrogen) are adsorbed epitaxially on some substrates, and therefore form a surface superstructure which influences further epitaxial growth.

The effect of contamination caused by cleaving of the substrate *in vacuo* $(10^{-4}-10^{-5}$ torr) on the epitaxial temperature is seen from Table 13. The metals concerned here are those growing epitaxially on a (100) surface of NaCl. It is obvious that the epitaxial temperature is lower for the substrate cleaved in vacuum $(V)$ than for analogous films prepared on air-cleaved crystals $(A)$.

Epitaxial Temperature of Films Prepared on Air-Cleaved (A)  *Table 13*
and Vacuum-Cleaved (V) (100) NaCl, Respectively

| Films | Au | Al | Ni | Cr | Fe | Cu | Ge |
|---|---|---|---|---|---|---|---|
| $T_e(A)$ (°C) | 400 | 440 | 370 | 500 | 500 | 300 | 500 |
| $T_e(V)$ (°C) | 200 | 300 | 200 | 300—350 | 300—350 | 100 | 350—400 |

The question arose whether further lowering of the epitaxial temperature might be achieved by preparing films on substrates cleaved in ultra-high vacuum ($\lesssim 10^{-9}$ torr). It has been found, however, that in epitaxial growth of Cu, Ag and Au on NaCl the epitaxy was inferior (but not for Al and Ni). Films on crystals cleaved in ultra-high vacuum are sometimes oriented differently from the normal, e.g. the (111) orientation appears instead of (100).

This fact suggests that some surface contamination is necessary for the formation of epitaxial film. It is supposed that in the case of epitaxy on NaCl, contamination consists in an adsorption of water that etches the crystal surface and effects a kind of recrystallization favorable for the growth of epitaxial film.

An additional factor which affects epitaxial growth is the electric field. It has been observed that growth is accelerated by this field for some semiconductors (Ge, Si, GaAs) prepared by a chemical trasnport reaction (see Sect. 2.1).

As we have noted earlier, systems with epitaxial films are very heterogenous as to chemical nature and structure, and are also prepared by various methods, the most important of which are vacuum evaporation, cathode sputtering and chemical vapor deposition. Because of the complexity of these phenomena and the presence of so many factors, no consistent and general theory has as yet been constructed. There are plenty of rules and hypotheses, most of which deal with special cases and do not cover the whole phenomena in their totality. Let us set forth here some of the theoretical ideas.

Brück and Engel [13], for example, assumed for epitaxial growth of metals on alkali halides that growth proceeds in such a way that the sum of the distances between the atoms (ions) of the metal and the halogen ions remains minimal. This corresponds to maximum coulomb forces in the system. The temperature dependence of epitaxy is explained as a consequence of the decrease in ionization energy of a metal atom adsorbed on the crystal surface compared with that of a free atom; this permits ionization to occur at relatively low temperatures. Epitaxy starts at a temperature at which the electrostatic interaction with the ionized metal is so strong that the particles can occupy equilibrium positions corresponding to an orientated structure. This theory only described the behavior of systems qualitatively because no quantitative information is known about the decrease in ionization energy. The model is in qualitative agreement with the experimental fact that the epitaxial temperature of metals on alkali halides increases with their ionization energy (e.g. in sequence Ag, Cu, Al and Au).

Another theory assumes that no ionization of atoms occurs on the surface; the atoms are only polarized and thus the effect of temperature operates through a change in atomic polarizability. According to another view epitaxy is realized through two directions with the closest-packed atomic arrangements in the salt being parallel to the two analogous directions in the metal.

Present methods of observation of film structures during their preparation (see Chapter 5) reveal that epitaxy begins to take place in the nucleation stage. Epitaxy should therefore be explained by nucleation theory. However, current theories are not yet developed sufficiently to manage this complex task. For the present they only enable certain relationships to be predicted in a qualitative way.

Thus, according to the capillarity theory of nucleation, it may be expected that for orientations with a lower free interfacial energy on the boundary between the substrate and film, the energy required for the formation of a critical nucleus will be lower so that a higher nucleation rate is to be expected for this orientation.

According to the atomistic model, epitaxy results from the arrangement in the smallest cluster which contains just one more particle than the critical nucleus. If supersaturation decreases, the size of the critical nucleus increases (one, two, three atoms, etc.). A minimum number of bonds exists in a three-atom embryo forming a [111] orientation (for an fcc crystal). The next cluster, of four atoms, forms a [100] orientation, and a five-atom cluster corresponds to a [110] orientation. A similar correspondence has been observed for epitaxial films of Au and Ag. These are, however, very special cases. As yet no general explanation can be advanced to account for even the simplest cases of epitaxy.

# CHAPTER 5

# COMPOSITION, MORPHOLOGY AND STRUCTURE OF THIN FILMS

Properties of thin films are determined by their chemical composition, the content and type of impurities in the bulk and on the surface, crystal structure of the bulk and surface, and the types and density of structural defects. Thus, all these parameters should be known and, if possible, modified during deposition.

## 5.1 Methods for Determination of Chemical Composition of Films

When looking for methods of chemical analysis of a thin film we are usually restricted by the small amount of substance which is at our disposal. Nevertheless for films that are not too thin, ordinary techniques of chemical analysis may be used, such as gravimetric and volumetric techniques and, especially, polarographic and chromatographic ones. A very sensitive technique is optical spectral analysis, which achieves spectrophotometric detection of substances for concentrations from $10^{-4}\%$.

Another very sensitive method is that using X-ray fluorescence, which has been mentioned already in connection with thickness measurements. The characteristic fluorescence of individual elements is excited by X-ray irradiation of the sample. The intensity of the particular wavelength emitted is then a measure of the concentration of the given element.

Concentrations down to $10^{-4}\%$ and a mass down to $10^{-13}$ g may be detected by X-ray microanalysis employing a sharply focused electron beam (diameter up to 2 μm, wavelength a few tenths of nm) for excitation.

The most sensitive are, of course, mass spectrometry methods, which are based on the identification of individual particles of a particular type and which make possible the detection of admixtures down to $10^{-7}\%$ concentrations. Ions entering the working chamber of a spectrometer may be liberated from the film and ionized in various ways, e.g. by spark discharge, laser pulse or ion bombardment. The latter technique, in particular, is of very high value (it is denoted in the literature by the abbreviation SIMS: Secondary Ion Mass Spectrometry). Ion-ion secondary emission, which is the underlying physical phenomenon, has been known for a long time (since 1938) but was considered unfit for surface analysis. Application of this phenomenon has been made possible by the use of extremely low energies of the primary ions and by the improved technology of ion beams and detection of low ion currents.

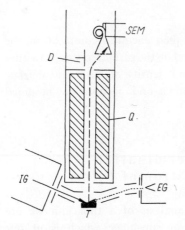

*Fig. 59a:* Experimental apparatus for SIMS: IG — ion gun; EG — electron gun; T — target; Q — quadrupole mass spectrometer; D — deflectionplate; SEM — channel electrom multiplier.

The apparatus most frequently used for this purpose is shown schematically in Fig. 59a. Monochromatic ions of an inert gas, usually argon, impinge on the target; the secondary ions pass through a diaphragm into the mass spectrometer, which is in this case represented by a quadrupole mass filter. The ions with a given mass (i.e. with a given specific charge $e/m$) are detected by a channel electron multiplier after deflection. The electron gun serves to compensate the positive charge, induced on surfaces of insulators by primary ions, by the negative charge of electrons.

The mechanism of ejection of secondary ions consists in a complex process of transfer of the momentum supplied by a primary ion to the atoms of the target (see Section 2.2.1). Besides the secondary ions there are also neutral particles ejected from the surface (roughly $3-30$ neutral particles per ion) together with electrons and photons. The yield is usually of the order of $1\%$ and depends on the energy and type of ion and on the material of the target. The yields vary considerably with the material. For example, the yield is 100 times higher for Al than for Au under bombardment by 12-keV argon ions. For a quantitative evaluation of spectra, it is necessary that the yields involved be known (up to the present, yields have been measured for about 35 pure elements). The impingement of primary ions on the surfaces produces not only atomic ions but also molecular ions and various radicals arising from decomposition of compounds. An example of a spectrum is shown in Fig. 59b. In principle both types of ions may arise. The materials with low ionization potential usually form positive ions, while those with

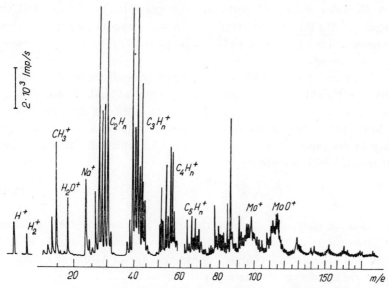

*Fig. 59b:* The mass spectrum of a contaminated Mo surface obtained by SIMS technique.

high potential usually form negative ions. The energies of emitted ions are $1-3$ orders of magnitude higher then the thermal energies at the boiling points (see also Sect. 2.2.1) and their distribution has a peak at about 10 eV and a half-width of $20-50$ eV. Since this dispersion is not too great, spectrometers with single focusing may be used.

The fact that the mean energy of the ions is of the order of a few eV means that the ions originate in practice only from the upper monolayer of the target. This renders the method very fit for the examination of surface structure. The bombardment gradually removes the upper monolayer. The removal time depends on the density of the primary current. The surface coverage varies with time as

$$\Theta(t) = \mathrm{e}^{-t/\tau} \tag{5.01}$$

(for $t = 0$, $\Theta = 1$) where

$$\tau = \frac{n_0}{v\gamma} \tag{5.02}$$

is the mean lifetime of the monolayer, $n_0$ is the density of particles in the intact monolayer, $v$ the density of the primary beam and $\gamma$ is the sputtering yield. In order to prevent damage to the examined specimen it is necessary to choose $v$ so small that $\tau$ is measured in hours. Sufficient sensitivity is obtained by bombarding the largest possible area and by using a detector which identifies individual particles. For example, the 3-keV Ar-bombardment of Mo removes less that 1% of a monolayer in 20 min. For some substances which have a high yield, the sensitivity reaches $10^{-6}$ of a monolayer (corresponding to $10^{-13}$ g/cm$^2$).

When the density of the primary beam is small, the method is called a static one (SSIMS). On the other hand, when a depth profile of some substance in a thin film is required, a method called the dynamic method (DSIMS), which uses higher flux densities, is employed. The removal of surface layers proceeds quickly. The rate with which the area examined shifts into the bulk is given by

$$x = \frac{M}{\varrho} \cdot \gamma \cdot \frac{I_p}{L \cdot e} \tag{5.03}$$

where $M$ is the molecular weight of the target, $\varrho$ its density, $\gamma$ the sputtering yield (atoms/ion), $I_p$ the primary flux density (A/cm$^2$), $L$ the Loschmidt number and $e$ the electron charge. For example, for an Ar-flux of 1 mA/cm$^2$, the removal rate for an iron target is 10 nm/sec.

If we are interested in the local distribution of some substance, we may use Secondary Ion Microprobe Mass Spectrometry (SIMMS). This technique employs a beam of primary ions which is sharply focused to a diameter $\lesssim 1$ µm and which scans the surface to be examined. The spectrometer is set for the mass in which we are interested and the signal is displayed on a screen or recorded on a photographic plate.

## 5.2 Electron Microscopy of Thin Films

### 5.2.1 Transmission Electron Microscopy

We have noted already that the progress in the development of electron microscopy represented one of the conditions for the development of the physics and technology of thin films. Its use enables the formation of thin films to be observed in situ together with bulk properties and surface structure. By combining various electron-microscope observation techniques, complete information on a given film may be obtained, especially in combination with electron diffraction methods.

The most important method used in the field of thin films is transmission electron microscopy.

An electron microscope is a relatively complex electron-optical system in which a focused electron beam is used to form an image of a given object and which uses a whole system of electron-optical lenses for magnification of the image.

The fact that an image may be formed by using a beam of material particles in a manner similar to the use of light rays in optics rests on the analogy between the fundamental principles of light propagation and of the motion of material particles. In practice it means that we can produce electric and magnetic fields of such a configuration that they will influence a beam of charged particles (namely, electrons) in a way that is analogous to the effect of normal lenses on beams of light. The analogy goes even further, the electron-optical lenses also have a cylindrical symmetry and the beam of electrons usually passes through the center of the system, i.e. along the optical axis.

Electrostatic lenses are systems of circular diaphragms or coaxial cylinders biased by suitable voltages; magnetic lenses are substantially coils fed by a current and usually placed inside a ferromagnetic shield equipped with polepieces to concentrate the magnetic field in a very small area.

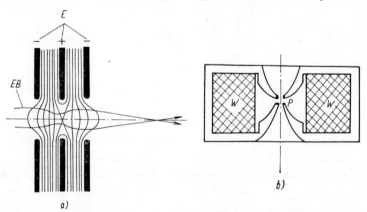

*Fig. 60a:* Electron-optical lenses: (a) electrostatic unipotential: E — electrodes, EB — electron beam; (b) magnetic with pole pieces: W — winding, P — polepieces.

Diagrams of one type of electrostatic lens (a unipotential lens) and one type of magnetic lens are shown in Fig 60a. The modern electron microscope generally makes use of magnetic lenses with which comparatively small focal lengths can be achieved that may be varied in a simple manner.

The principle of the electron microscope will be explained here only briefly. (More detailed information may be found in [2].)

The basic diagram of the electron microscope is given in Fig. 60b. Electrons are emitted from a heated cathode surrounded by a negative electrode (Wehnelt's cylinder) and are accelerated by an anode. The first electron lens — the condenser — focuses the electron beam on the sample. The objective lens is close to and under the sample; it produces the first magnified image of the sample. The image is further magnified by the projection lens and is projected onto a fluorescent screen or photographic plate. Actually, microscope systems are usually somewhat more complex and use more lenses than the objective-projector system described here. The more complex arrangements enable the magnification to be varied over a greater range. In the two-stage system a magnification of $5000-25000$ is commonly achieved whereas more elaborate systems may have a ratio of magnification of 1 : 50 and the direct magnification of 250 000.

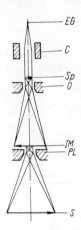

*Fig. 60b:* Schematic of transmission electron microscope: EG — electron gun; C — condenser lens; Sp — specimen; O — objective lens; IM — intermediate image; PL — projection lens; S — screen.

The magnification is not the only important parameter of the microscope. The resolving power of the microscope is also of great importance. It is well known that the resolving power in optical systems is limited by diffraction effects, as a result of which an image of a point is a kind of

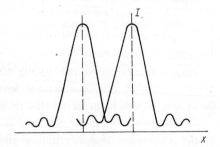

*Fig. 61:* Diffraction patterns produced by two close points.

diffraction pattern with a periodic distribution of light intensity (Fig. 61). Two points can be distinguished from each other if the interference maxima are removed from each other by at least a halfwidth of the main maximum.

Electrons exhibit in fact, a wavelike nature and so a definite wavelength may be ascribed to them which is given by wave mechanics as

$$\lambda = \frac{h}{p} \tag{5.1}$$

where $h$ is Planck's constant and $p$ is the momentum of the electron. Thus the wavelength depends on the velocity of the electron. For example, for electrons accelerated by a 50-kV voltage the wavelength is 0.005 nm.

The resolving power $\delta$ is usually given by

$$\delta \sim 0.61 \frac{\lambda}{\sin \alpha} \tag{5.2}$$

where $\alpha$ is the aperture half-angle, i.e. the angle at which the electrons enter the system. This angle is usually very small, of the order of $10^{-2}$ to $10^{-3}$ radian. The theoretical resolution of the electron microscope would be then about $0.2 - 0.3$ nm. In fact it is further limited by the optical aberrations of the system, for as with light optics, aberrations of lenses also exist in electron microscopy. The most important aberrations which come into play in an electron microscope are chromatic and spherical aberrations and axial astigmatism. Chromatic aberration has its source in the fact that not all the electrons have precisely equal velocities so that they are focused at different distances from the lens (as in optics where different wavelengths correspond to different focal lengths). The term spherical aberration denotes the fact that the beams proceeding along directions far from the optical axis do not focus on the focal plane parallel to the plane of the lens but on a spherical surface. The aberration increases with the cube of the distance from the axis and is reduced by selecting very thin paraxial beams, i.e. small aperture angles. The optimum situation occurs when spherical aberration just equals the blur caused by diffraction; this leads to a resolution of $\sim 0.28$ nm.

This is therefore the theoretical limit of resolution. Owing to chromatic aberration and axial astigmatism the value deteriorates to about 0.4 to 0.5 nm, a resolution actually attained by the modern electron microscope.

An important question is how contrast is obtained in the electron microscope. In the light microscope contrast arises from the different absorption of light in different parts of the sample. In the electron microscope the situation is different. The contrast results here chiefly from differences in electron scattering. Although absorption of electrons also takes place it is not so significant and is moreover undesirable as it leads to heating of the sample.

The samples used in transmission electron microscopy have to be sufficiently thin not only to avoid high absorption of transmitted electrons but also to avoid multiple scattering of electrons, which would be detrimental to the resolution. It may be said that the resolution cannot be better than 1/10, and on ultrathin films 1/20, of the film thickness.

Samples are therefore usually in the form of films with thicknesses ranging from tens to hundreds of nm. Some films of these thicknesses may be prepared as self-supporting ones. In other cases (especially on observation of ultrathin films) suitable substrates have to be used. These must be without structure of their own and as transparent as possible for electrons (i.e. of low atomic number). Such a substance is carbon, which may be prepared by evaporation in the form of an ultrathin self-supporting film, is amorphous, chemically inert, can withstand heating up to 200 °C and does not alter under electron bombardment. Another possibility is to use layers of organic materials such as celluloid, $Al_2O_3$ prepared by anodic oxidation, glass, etc.

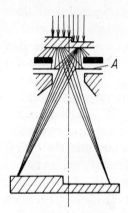

*Fig. 62:* Origin of contrast due to scattering.

For some purposes (e.g. the study of epitaxy) it is necessary to use a substrate with a crystalline structure. Mica, $MoS_2$, MgO, etc. may be used in such cases.

As noted above, contrast arises because of the scattering or diffraction of electrons. The origin of the scattering contrast is shown schematically in Fig. 62. The region of greater thickness (or the region containing a substance with higher atomic number) scatters the electrons by a larger angle. The aperture A prevents the passage of the more strongly deflected beams so that the regions of stronger scattering appear darker.

The diffraction of electrons is important for the study of crystalline substances. The phenomenon of diffraction depends on the spacing of crystal planes $(d_{hkl})$ and their orientation $\vartheta$ with respect to the incident beam. The Bragg condition for a diffraction maximum is

$$2d_{hkl} \sin \vartheta = n\lambda \qquad (5.3)$$

where $n$ is an integer. The typical value for fcc metals is $d \approx 0.20$ nm, which for $\lambda = 0.004$ nm gives $\vartheta = 10^{-2}$ rad and so the diffracted beam misses the centre of the aperture. If crystals are variously oriented with respect to the electron beam, the corresponding spots on the screen will be of different brightness (Fig. 63). The diffraction contrast also enables the lattice defects to be observed.

The existence of the diffraction contrast enables in certain cases, the image of the crystal lattice to be formed directly. If the lattice constant $d$ is large enough for the diffracted beams to pass through the aperture under

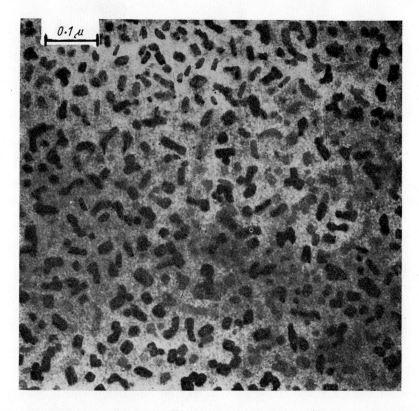

*Fig. 63:* Contrast produced by different crystal orientations (LiF on carbon substrate, t = 2 nm (Anderson)).

certain conditions, an interference pattern may arise which will correspond to the structure of the lattice, and provided the lattice constant is at least 0.4 nm, a direct image of the lattice may be obtained. For the smaller lattice constants, a beam of very fast electrons of oblique incidence has to be used. In this way, a resolution below 0.1 nm may be attained.

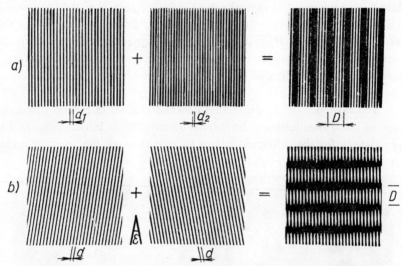

*Fig. 64:* The optical analog of moiré pattern formation: (a) mismatch: (b) rotation.

A very important technique for the study of a crystal lattice and its defects makes use of the double diffraction of two overlapping lattices, known as the moiré method. The essence of this technique can best be seen from an

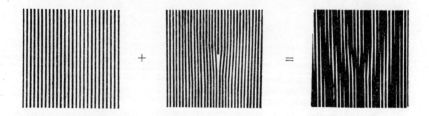

*Fig. 65:* The optical analog of the formation of a dislocation moiré pattern.

optical analogy. If two periodic structures consisting of alternate bright and dark fringes are laid upon each other, two possible types of moiré patterns may be formed (Fig. 64). In the first case the crystal lattices run parallel but they have different lattice constants $d_1$ and $d_2$. In this case the parallel moiré fringes are formed with a spacing

$$D = \frac{d_1 d_2}{d_1 - d_2}$$ (5.4)

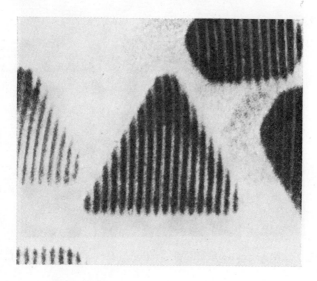

*Fig. 66:* Moiré patterns (Au on $MoS_2$ (Jacobs)).

In the second case both lattices have the same constant $d$ but they are rotated with respect to each other by a small angle $\varepsilon$. The ensuing moiré fringes have a spacing $D = d/\varepsilon$. In the general case of different lattice constants and rotation angle $\varepsilon$ the spacing is given as

$$D = \frac{d_1 d_2}{\sqrt{(d_1^2 + d_2^2 - 2d_1 d_2 \cos \varepsilon)}}$$ (5.5)

A magnification of the lattice constant $d_1$ is thus realized, its value $m$ being given as

$$m = \frac{d_2}{\sqrt{(d_1^2 + d_2^2 - 2d_1 d_2 \cos \varepsilon)}}$$ (5.6)

By this technique various lattice defects may be made visible, as may be seen, in the case of dislocation, by the optical analogy in Fig. 65.

An electron-optical illustration of thin crystallites of gold on a $MoS_2$ substrate with parallel moiré fringes is shown in Fig. 66 and a film with dislocations (the same system) is illustrated in Fig. 67.

In addition to the method described for the observation of objects in a transmission microscope, several additional, more specialized techniques exist, among which is the method of dark-field images.

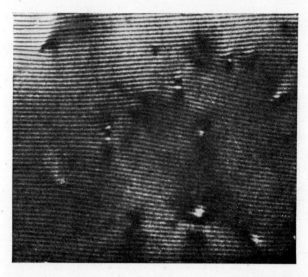

*Fig. 67:* Moiré pattern with dislocations (Au on MoS$_2$ (Jacobs)).

The principle is again analogous to that of observation in a dark field used in light microscopy. While in the normal mode of observation the diffracted beam is screened out and a direct beam is used for image formation, the opposite process is adopted here. As can be seen from Fig. 68,

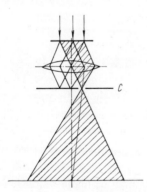

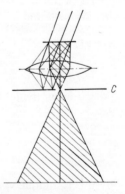

*Fig. 68:* Image formation in dark-field mode (the aperture C is shifted from the optical axis of the system).

*Fig. 69:* Alternative possibility of image formation in dark-field mode (the aperture C is well centered, but the beam arrives at an oblique angle).

the beams passing directly through the system are screened out by the aperture C placed asymmetrically with respect to the system axis, while the diffracted beam is allowed to pass through it. The arrangement described has the drawback of using beams far from the optical axis. The image exhibits a large spherical aberration detrimental to the resolution. Therefore the system shown in Fig. 69 is used instead. Here the electron beam is

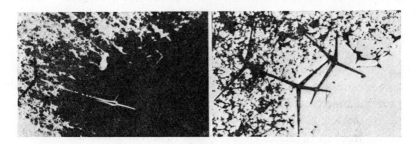

*Fig. 70:* The comparison of bright- and dark-field images (ZnO crystals).

incident on a sample at a such an angle that the diffracted beam which we wish to use for image formation proceeds through the centre of the system. In this way better resolution is achieved but it is necessary to tilt the optical system and this is technically a rather exacting operation. A comparison between an image in the bright-field and dark-field mode is given in Fig. 70.

In transmission electron microscopy use is also made of other techniques such as shadow microscopy and interference microscopy, etc. Shadow microscopy is specially suitable for the observation of magnetic microfields in a sample.

An important application of transmission electron microscopy is in the observation of ultrathin films, especially during their growth, as has been mentioned earlier. It should be added here that the deposition of film must proceed directly in the viewing field of a microscope, i.e. the microscope has to be equipped accordingly.

It is necessary to keep in view the fact that with the relatively low vacuum frequently used in microscopes $(10^{-4} - 10^{-5}$ torr$)$, contamination of the surface occurs in the course of observation; this may, especially on investigation of nucleation processes, markedly influence the growth of film and its properties. Contamination arises from hydrocarbons from the pumping fluid of the diffusion pump. The adsorbed layer of organic substance disintegrates upon the impingement of the electron beam and forms a polymer, whose thickness increases with the duration of electron irradiation. It has been found that under conditions common in electron microscopes, the thickness of the layer increases at a rate of 0.05 – 0.1 nm per second. In order to eliminate or at least to check this effect, microscopes designed for the study of nucleation are usually equipped with a special chamber enclosing the sample together with a deposition apparatus which is separately exhausted by, for example, a titanum sublimation pump. Since the chamber is only connected with the remaining space of the microscope by two small apertures, it is possible to reduce considerably the pressure in the chamber and lower the content of hydrocarbons in the residual gas considerably.

The electron beam affects the process of film formation even in a completely clean environment because it represents a certain amount of electrical charge. As we have seen in Chapter 4, the surface charge may influence both the nucleation proper and the subsequent process of island coalescence. We should therefore be careful in interpreting results. A further effect of the electron beam is heating of the film owing to absorption of a certain fraction of the electrons by the film. The effect depends on the density of the electron beam and is especially significant in thicker samples.

### 5.2.2 Electron-microscopic Examination of Surfaces by Replica Method

In some cases we are interested in the surface structure of a film. The electron-microscopic methods are well suited for this purpose (in addition to the optical interference methods mentioned in Sect. 3.3.2 and the LEED method dealt with in Sect. 5.3.2).

The replica method is the one most frequently used. According to the nature of the sample, either the one-stage or two-stage replica technique

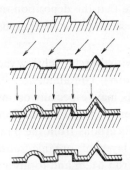

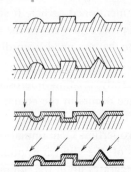

*Fig. 71:* The processing steps in formation of a one-stage replica.

*Fig. 72:* The processing steps in formation of a two-stage replica.

is used. If the sample can be dissolved, the one-stage technique is applied. The procedure is shown in Fig. 71. A thin layer of shadow material is evaporated at a large angle of incidence onto the surface to be examined. For the shadow material, a metal with a high atomic number and high melting point is usually used, e.g. platinum. In this procedure the film is covered by a layer of varying thickness with a material that strongly scatters electrons and in such a way that the layer is thick on the areas oriented perpendicular to the direction of incidence, whereas shadows are produced behind irregularities; the shadow lengths depend on the incidence angle and height of the irregularity. On the film thus processed, a carbon layer is deposited at normal incidence and with sufficient thickness to be self-supporting. The original film is then dissolved in a suitable solvent and the shadowed carbon foil is examined in the microscope.

If the film cannot be removed, the two-stage technique has to be used. The procedure is shown in Fig. 72. An impression thicker than the film is first made and is then peeled mechanically from the film. For this purpose a solution of collodion in amyl acetate or formvar dissolved in ethylene dichloride may be used; the solution is deposited on the surface and the solvent left to evaporate, or a polystyrene impression is made. After removal from the specimen, the imprint of the specimen surface is coated with a thin layer of carbon which is then freed by dissolving the collodion or polystyrene matrix. The replica is scooped from the solvent with a sieve that is used as a support in a microscope and the contrast is usually enhanced by oblique

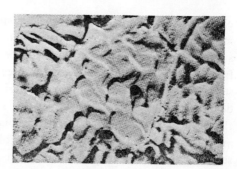

Fig. 73: The shadowed replica of a surface of etched copper.

deposition of a shadowing material as described above. The resolution is, however, higher on one-stage replicas. The value attainable is about 2 nm. The resolution is limited by the granularity of the shadowing material which is the reason for using a material with a high melting point such as platinum or palladium.

An example of the replica image of a surface is given in Fig. 73.

### 5.2.3 Special Types of Electron Microscopes for Direct Image-forming of Film Surface

#### 5.2.31 Scanning Microscope

The principle of the scanning electron microscope rests on a periodic scanning of the surface by a focused beam deflected with the help of a raster similar to that of television. The image has to be displayed on a screen with a long afterglow. If the microscope is constructed for the transmission mode it has the advantage that the specimen is much less affected by the

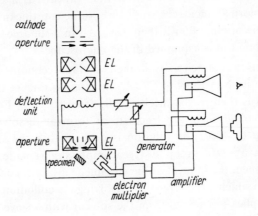

*Fig. 74:* Schematic of a scanning electron microscope.

electron beam than in the classical transmission microscope. However this microscope may be also constructed to form an image of the specimen surface. A schematic illustration is given in Fig. 74. A beam of electrons from an electron gun is focused by three electron lenses EL. Between the second and third lens, the beam is periodically deflected by a system of deflection coils so that it runs over the surface under examination. When the beam impinges on the surface, secondary electrons are emitted which are extracted by an electric field to the collector $K$. The intensity of current recorded by the collector is determined at every instant by the emission properties of the area of impingement involved. The signal is usually amplified by an electron multiplier and amplifier connected in series. The signal determines the brightness of the spot on the screen of the cathode-ray tube whose raster is synchronized with that of the microscope. On the screen of the tube (again, with a long afterglow) an image of the surface of the specimen is formed which corresponds substantially to the distribution of the coefficient of secondary emission and to the geometrical shape of the surface, which in turn modifies the shape of the electric field close to the surface. The contrast is also affected by local electric or magnetic micro-fields. (The contrast is therefore caused by other physical factors than those in transmission electron microscopy.)

The resolution of the apparatus is limited first by the fact that the diameter of the scanning electron beam cannot be arbitrarily diminished, and also by the fact that the secondary electrons are always created in a certain finite region around the site of impingement. To achieve the contrast, it is necessary that a sufficient number of primary electrons strike the specimen so that the actual differences in the secondary emission are not obscured by statistical fluctuations of the current. Thus the higher the current density of the beam and the shorter the probing time (i.e. the time during which the beam is focused on a given spot), the smaller is the detail which can still be observed.

The limit of resolution is about 20 nm, and therefore lower than that of the transmission microscope. But, in a contrast to that microscope, it possesses far larger depth of focus. (This feature has, however, more importance in other fields, e.g. biology.) Its magnification can also be varied over a broad range (i.e. from 20 to 50 000), while the specimen is affected very little by the electron beam. In recent years the scanning microscope has been widely used, especially in the examination and checking of microcircuits.

### 5.2.32 Reflection Microscope

The principle of the reflection microscope is illustrated schematically in Fig. 75. The electrons from the gun are deflected by a magnetic field $M$ and arrive at the object, which constitutes part of the electron mirror.

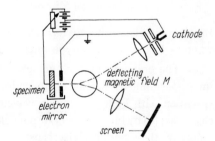

Fig. 75: Schematic of a reflection electron microscope.

A charged particle can surmount only a decelerating potential that is lower than that corresponding to its kinetic energy. If the particle reaches the equipotential line at which its kinetic energy is zero (the particle has been stopped completely), it rebounds from the line and begins to return. The returning electrons are now deflected to the other side in the region of the field $M$, and after passage through a projective lens they produce an image on a fluorescent screen.

The equipotential line from which the electrons are reflected runs close to the surface of the specimen so its shape is determined by the microgeometry of the specimen on the one hand and on the other by local variations in potential on the specimen caused by variations in the work function of individual sites (i.e. contact differences of potential). In this way the electrons impinging perpendicularly obtain different tangential components of velocity, giving rise to the contrast between the sites.

The electrons do not hit the specimen in this mode of image-forming and so cause neither heating nor any other undesirable effects.

The theoretical limit of the resolving power is 10 nm at an accelerating voltage of 40 kV. In practice, the resolving power is less because the impinging electrons are not strictly monoenergetic. Good resolution is achieved in the vertical direction, i.e. the resolution of the heights of two adjacent sites, and is about 2 nm. If the contrast is due to variations in potential, then the smallest observable potential difference is about 0.03 eV.

A scanning version of this microscope has also been developed. The contrast is then produced somewhat differently. The potential of the specimen is chosen so that some electrons from the beam, which is not strictly monoenergetic, are reflected from an equipotential line and are recorded; others impinge on the specimen and are absorbed by it (their energy upon impingement is so small that secondary emission is negligible). Those areas will therefore appear dark.

The recording system is similar to that of the scanning microscope.

### 5.2.33 Emission Microscopes

In the transmission microscope the source of electrons is the thermocathode whereas in the emission-type microscope the source is the specimen itself.

The emission may be achieved (a) by heating the specimen; this is image-forming by means of thermionic electrons; (b) by irradiating the specimen with light of suitable wavelength; in this case we obtain photoelectrons; (c) by impingement of particles (e.g. electrons), which produces the image by the secondary electrons.

In all these cases the electrons emitted from the sample are very slow (their energies being usually equal to fractions or units of an eV) and they have at first to be accelerated. The acceleration, together with the first focusing, is provided by a cathode lens, which is a very important component of the emission microscope (Fig. 76). A narrow aperture, Ap, is situated in the focal plane and has a function similar to that in the transmission microscope. A lens (or a lens system) is placed to form the magnified image. The resolution of this type of microscope depends on the energies of the electrons emitted from the specimen and on the intensity of the electric field at the specimen surface. The higher the initial energy of the electrons and the lower the intensity of the field, the smaller the resolving power. The theoretical limit of the resolving power is about 10 nm for thermionic emission and photoemission; it is less for secondary emission due to higher initial velocities of the electrons. In practice, resolutions of several tens of nm are attained.

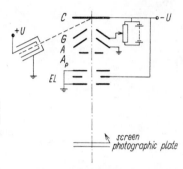

*Fig. 76:* Schematic of an emission microscope; U — potential; C — cathode; G — grid; A — anode; $A_p$ — aperture; EL — electrostatic lens.

For the examination of the surface by the secondary electrons, ions are frequently used as the primary particles. The secondary electrons then originate from the surface of the specimen whereas when using electrons as the primary particles the secondary ones arise from a certain depth below the surface. Each method therefore provides different information about the specimen.

Besides variations in the coefficient of secondary emission, another factor plays a role in producing the resultant contrast, namely, the geometrical relief of the surface, which modifies the shape of the potential near the surface, thus determining the direction of the initial path of the secondary electrons. As the primary particles impinge obliquely, the shadow effect also makes a contribution. Accurate interpretation is therefore very difficult.

### 5.2.4 Tunnel Emission and Field Ionization

The tunnel effect is a quantum mechanical type of particle transfer through a potential barrier. Fig. 77 illustrates this effect, which is used in one type of special electron microscope. The figure illustrates the energy diagram of the metal-vacuum interface, when an electric field of intensity $F$ is applied perpendicularly to the surface. ($E_c$ is the bottom level of the conduction band, and $E_a$ is the potential energy of electrons in vacuum if $F = 0$.) At

134

moderate temperatures, the electrons in the metal occupy levels up to the Fermi energy, and to enable them to overcome the surface barrier and emerge in the vacuum, an energy has to be supplied to them at least equal to the work function $\chi_0$. If the magnitude of the field is $F$, the potential decreases outside the metal as $eFx$ (curve 1). If we also take into account the existence of the image forces $e^2/4x$, the lowering of the barrier caused by the field $F$ will assume the form illustrated by curve 2. If the width of the barrier at the energy level corresponding to the highest occupied states is sufficiently

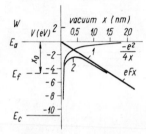

*Fig. 77:* Energy level diagram of metal (W)-vacuum interface in the presence of a strong electric field.

narow, i.e. the field is strong enough, then according to quantum mechanics there is a probability that an electron will pass through the barrier by means of the tunnel effect (i.e. without attaining or surpassing the energy of the barrier peak). Under these conditions an emission current may be observed, the density of which, according to quantum mechanical calculations and under certain simplifying assumptions, is

$$i = BF^2 \exp\left(-\frac{C\chi_0^{3/2}}{F}\right) \qquad (5.7)$$

where $B$ and $C$ are constants depending on the nature of the cathode. This is the so-called Fowler-Nordheim's equation, from which it is evident that besides a strong dependence on the field intensity $F$, the current depends on the work function of the given surface. The emission currents of practical applicability correspond at normal values of the work function $(4-5\ \text{eV})$ to a field intensity $5 . 10^8 - 5 . 10^9\ \text{V/m}$.

On the surface of a metal which is set at a high positive potential, an analogous effect occurs. If a neutral atom approaches within a short distance of the surface, the valence electron of the atom may, through the tunnel effect, go over to a free electron level in the metal; the atom is thus ionized and, as an ion, is strongly repelled by the electric field from the metal surface, i.e. emitted from the vicinity of the negative electrode.

### 5.2.41 Field Electron Microscope

The tunnel emission of electrons has been employed in the field electron microscope constructed by E. W. Müller. A diagram of it is shown in Fig. 78.

A cathode, $C$, from the examined material is made in the form of a very delicate tip (with radius curvature $\leq 1\ \mu m$), which is mounted on a metal support, $B$, used for heating the tip to degas and clean its surface.

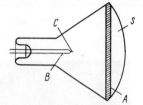

*Fig. 78:* Field electron microscope: C — cathode; B — support (heater) A — anode; S — screen.

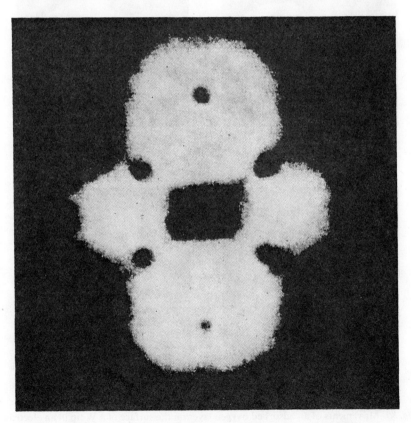

*Fig. 79:* The image of a W tip produced by the field-emission microscope.

In the evacuated tube an anode, *A*, in the form of a ring, is often placed directly on the wall of the bulb, a fluorescent screen, *S*, is also placed on the bulb wall. By applying a voltage of the order of several kV to the tip, a field arises of sufficient intensity to produce tunnel emission of adequate current density. Since the point of the tip consists usually of a single crystal (the size of the tip is comparable to that of a microcrystallite in a poly-crystalline material) which is rounded off (usually by electrochemical etching), different points of the surface correspond to different crystal planes and have different values for the work function. Owing to the strong depen-dence of the emission current density on the work function, different points of the tip surface will emit differently. The field at the tip surface is radial in

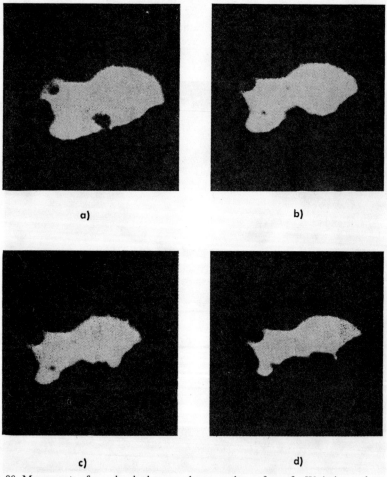

a)  b)

c)  d)

*Fig. 80:* Movements of an adsorbed oxygen layer on the surface of a W tip in an electron projector.

direction and thus emitted electrons will proceed along straight lines to the screen where they will form an image of the work function distribution over the tip surface. Thus a magnified emission image of the cathode (Fig. 79) will appear on the screen, unaffected by electron-optical aberrations, and the magnification will be approximately equal to the ratio of the screen and tip radii. It is then readily understood why magnifications of the order of $10^5 - 10^6$ may be achieved with ease. The resolving power in this case is limited mainly by the effect of diffraction of the electrons and is about 2 nm, in some cases even less.

This basically simple system is complicated by the fact that because of the strong dependence of the emission current on the work function, it is very sensitive to the adsorption of even slight amounts of residual gas and hence all measurements have to be carried out in ultra-high vacuum.

On the other hand, dependence of the current on the adsorption of thin films on the surface makes the method useful for the examination of thin films in the initial stages (see Sect. 4.2.4) and also for observing the surface migration of adsorbed films over the surface. In the former case, one may observe even very small clusters and their growth. In the latter case, the procedure consists in the evaporation of some substance onto the tip so that only a part of it is covered by the film. The boundary of the film is easily visible on the screen and its movement or its disappearance may be observed under various conditions especially various temperatures. An example of such a surface is given in Fig. 80. In this way, the coefficients of the surface diffusion and desorption may be determined, and these play

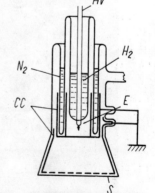

Fig. 81: Field-ion microscope: HV — high-voltage supply; $N_2$ — liquid nitrogen; $H_2$ — liquid hydrogen (or helium); CC — conducting coating; E — tip emitter; S — screen.

a very significant role in film growth as we have seen in Chapter 4. It is, however, necessary to take into account the presence of a very strong field at the tip surface which apparently may influence the surface processes substantially.

### 5.2.42 Field Ion Microscope

We should also mention the field ion microscope, a very important apparatus for the examination of the crystal structure of a surface. Such an apparatus is of great value in thin film physics, especially in the study of nucleation and epitaxy.

The field ion microscope employs the field ionization effect, which is briefly described on page 134. Helium atoms are mostly used for the image-forming. The apparatus does not differ much from the field electron microscope (Fig. 81). Since the intensity of the field required (of the order of $10^{10}$ V/m) is higher here, the radius of the tip is still smaller ($r \sim 100$ nm) and the voltage applied is of the order of 10 kV. The probability of ionization of helium atoms at the tip surface is markedly affected by the local field at the site of ionization, which in turn depends on the position relative to individual points in the lattice. The resolving power of the apparatus is not limited by the diffraction effect (because of the much greater mass of the ions and the consequently shorter wavelengths of the particles) but it may be limited by the thermal velocities of the particles. The tip and walls of the

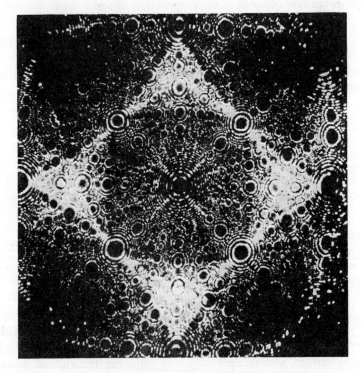

*Fig. 82:* Image of a Pt tip produced by an ion projector (E. W. Müller).

tube are therefore cooled. If the cooling is effected by liquid helium, the resolving power attains a value of $0.2-0.3$ nm, which is in the region of atomic dimensions. In these circumstances, the light spots which may actually be distinguished on the emission image correspond to individual atoms in the crystal lattice (Fig. 82). The apparatus would be ideal for the study of nucleation because it enables embryos of atomic dimensions to be observed. Unfortunately, at the intensities of electric field needed to produce a sufficiently sharp image, the mechanical forces acting on the tip surface are so large that adsorption is mostly impossible.

## 5.3 Diffraction of Electrons

We have pointed out already when dealing with the methods of electron microscopy that the passage of electrons through a solid is accompanied by their diffraction. The fundamental equation for the diffraction maxi-

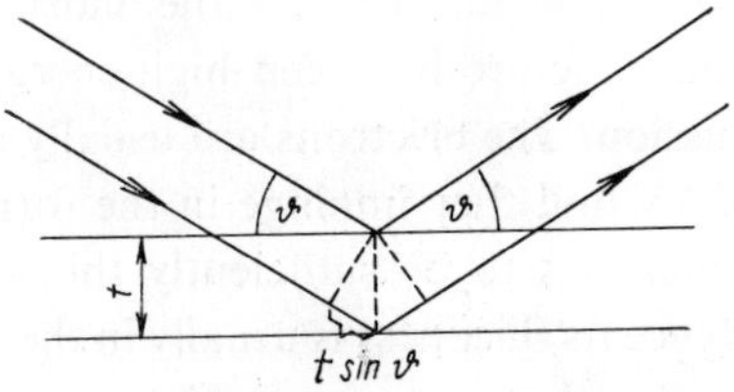

*Fig. 83:* Schematic illustration for the derivation of the Bragg condition.

mum — the Bragg condition — has been given in equation (5.3), which may be derived in a simplified manner from the model shown in Fig. 83. The scattered waves interfere with each other and the maximum intensity is achieved when their phase difference is equal to an integral multiple of $2\pi$, or their path difference (i.e. $2d \sin \vartheta$) is equal to an integer multiple of the wavelength $\lambda$. This model, while assuming a kind of mirror reflection on crystal planes, agrees with the results of the more exact theory of Laue, which considers diffraction of waves at the individual points of a crystal lattice.

For a complete analysis of a diffraction pattern it is not only the positions of the particular reflection spots that are important but also their intensity. The intensity is determined by $(a)$ the atomic scattering factor, which depends on the individual interactions of the electrons with particular atoms and is thus a function of the properties of the atoms (of the number and distribution of electrons in them), $(b)$ a structural factor which depends on the configuration of the primitive cell.

The theoretical treatment of these problems is very complicated and, as far as electron diffraction is concerned, not yet fully completed, so the exact interpretation of electron diffraction patterns is at times very difficult and sometimes impossible. It is assumed in 'kinematic' theory that electrons are not absorbed, no multiple scattering occurs (the crystal is not fully oriented for Bragg reflection), and that the impinging beam is mono-chromatic. Because the diffraction angles are small ($\leq 2°$) the 'column' model is used: the sample is divided into columns parallel to the primary beam and the intensity of the diffracted beam is calculated without inclusion of the interaction of the beams from neighboring columns.

A better elaborated but more complicated theory is a dynamic one which considers multiple diffraction and applies quantum mechanical methods to the problem. For more detailed information about both theories the reader is referred to the special literature [32], [41].

### 5.3.1 Diffraction of High-Energy Electrons in Transmission and in Reflection

($a$) In thin film physics the most frequently used method for the study of film structure has been high-energy electron diffraction (HEED), in trans-mission. The electrons are usually accelerated by a potential of several tens of kV and they impinge in the form of a focused beam onto the specimen, which has to be sufficiently thin so that electrons can easily penetrate it. Hence its thickness is usually in the region of hundreds of nm or less (depend-ing on the sort of material examined). The diffracted electrons arrive at a screen or photographic plate, where they give rise to a diffraction pattern (Fig. 84). It is also possible to obtain diffraction patterns on specimens of

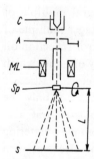

*Fig. 84:* Schematic diagram of electron-diffraction apparatus: C — cathode; A — anode; ML — magnetic lens; Sp — speci-men; S — screen.

much less thickness that are not self-supporting. In such cases, the film under study is deposited on an amorphous substrate. The diffraction pattern may be observed even on very small amounts of material, of the order of $10^{-12}$ g, and it is possible to carry out an analysis of very small regions and observe, for example, the structure of individual islands of growing film. An advantage of the technique lies in the short exposure times (seconds, in contrast to the hours-long exposures of X-ray diffraction).

The analysis of the diffraction pattern may yield information on the type of crystal lattice and interplanar spacings of the substance examined. If we denote the distance of the specimen from the screen by $L$ and the distance of the relevant spot in the diffraction pattern from the spot of the primary (undiffracted) electrons by $r$, the fundamental Bragg equation may be rewritten (by development of the sine function into a series) as

$$d_{hkl} = \frac{\lambda}{2 \sin \vartheta} = \lambda \frac{L}{r} \left[ 1 + \frac{3}{8} \left( \frac{r}{L} \right)^2 - \dots \right] \doteq \lambda \frac{L}{r} \qquad (5.8)$$

There are several possible types of diffraction pattern and it may be inferred at a glance to which type, as regards crystal structure, the given substance belongs. If the specimen is monocrystalline, a point diagram arises (see Fig. 85). Analysis of the positions of spots (see equation (5.8)), together with their intensities, may provide information on the crystal lattice. If the material is polycrystalline, i.e. the individual crystallites are variously oriented with respect to the incident beam, a pattern of concentric

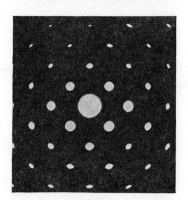

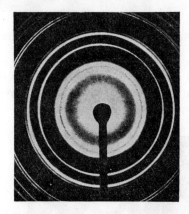

Fig. 85: Transmission electron-diffraction pattern of monocrystalline Cr film.

Fig. 86: Transmission electron-diffraction pattern of polycrystalline Ni film.

rings arises (Fig. 86) in which the radii of the rings correspond to the distances of the individual spots from the primary spot in a single-crystal pattern. The intensity of a particular ring is also measured (e.g. by photometric techniques) in this case. If the film is partially oriented, e.g. it has texture, the rings in the pattern are discontinuous (Fig. 87) and some rings may be missing altogether. Finally, for amorphous substances, the diffraction pattern is totally diffuse or consists of two or three diffuse rings. If the specimen is a very perfect single crystal, Kikuchi patterns, consisting of bright and dark lines, appear on the diffractogram. The lines result from

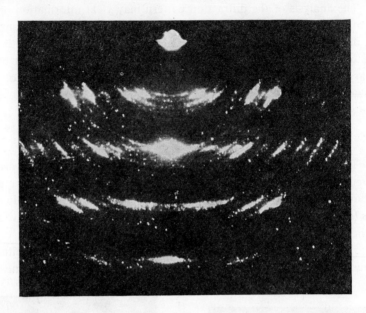

*Fig. 87:* Transmission electron-diffraction pattern of GaAs film with fiber texture.

a multiple diffraction of electrons in the crystal and their pronounced attenuation or transmittance in particular directions. The more perfect the monocrystal and its surface the more complex are the Kikuchi lines.

*Fig. 88:* Diffraction pattern with satellites (epitaxial Ag (100) film on NaCl; the extra reflections are due to twins (t) and double diffraction (d)).

Extra reflections or satellites are frequently observed in diffractograms, i.e. spots which do not belong to the basic pattern. These are usually situated symmetrically around the primary spot. They are caused by a special feature of diffraction in very thin crystal structures. Further, they may be a consequence of a multiple diffraction (a once-diffracted beam becomes a primary beam for another diffraction), which may also lead to modification of a spot's intensity and to a rise of 'forbidden' reflections. Another cause of the occurrence of such reflections is the presence of crystal twins. An example is given in Fig. 88.

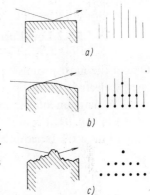

*Fig. 89:* Formation of RHEED patterns: (a) atomically smooth surface yielding streak pattern; (b) a degree of surface roughness generally leads to the appearance of spots and lines; (c) surface with large irregularities yields spot pattern.

(*b*) For the examination of the structure of surface layers, HEED in reflection is used (RHEED). The electron beam impinges onto the sample at a very small angle ($\sim 1°$) and has an energy of $30-100$ keV. Diffraction patterns differ according to the quality of the surface. If the surface is atomically smooth, the diffraction pattern consists of parallel lines (Fig. 89a); if it is not completely smooth, pronounced points appear on the lines (Fig. 89b) and, finally, if the surface is very rough, no proper 'reflection' occurs and the electrons pass through the irregularites and a transmission pattern arises which is made up of individual points (Fig. 89c).

*Fig. 90:* An electron reflection-diffraction pattern (Mo film with fiber texture (Chopra)).

The method of electron diffraction in reflection is less accurate than the transmission variant because the quantity $L$ in equation (5.8) cannot be determined so accurately. On insulating specimens, spurious effects are produced by the charging of the surface, which can be removed by simultaneous irradiation by low-energy electrons ($E \approx 200$ eV). The diffraction pattern depends also on the orientation of the specimen with respect to the impinging beam and the apparatus is usually equipped with a device for rotation of the pecimen in vacuum. An example of a reflection diffractogram is shown in Fig. 90. The pattern is for a surface film of $2-3$ nm thickness.

### 5.3.2 Low-Energy Electron Diffraction (LEED)

One of the most promising methods for the examination of a surface is low-energy electron diffraction, abbreviated as LEED.

A diagram of the experimental apparatus used for this purpose is shown in Fig. 91. From the electron gun Gu, consisting of a thermocathode with a low work function and focusing electrodes, which provide a focused

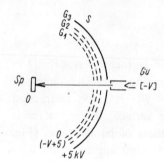

*Fig. 91:* Schematic diagram of LEED apparatus: Sp — specimen; Gu — gun; S — screen; $G_1$, $G_2$, $G_3$ — grids.

beam of low-energy electrons $(5-500 \text{ eV})$, the electrons impinge on the surface of the specimen, $Sp$. The diffracted electrons penetrate through three grids $G_1$, $G_2$, $G_3$ of large transmissivity $(\sim 85\%)$ and arrive at a collector covered with a luminophore (screen $S$) on which they produce a diffraction pattern that may be observed visually or photographed. To achieve a sufficiently intense pattern, it is necessary to use high accelerating potentials $(\sim 4 \text{ kV})$ between the specimen and collector. The grid $G_3$ electrostatically screens both impinging and diffracted electrons from the accelerating field. The grid $G_1$ usually has the potential of the specimen and is intended to create a field-free space around it (in some cases it is biased positively by a small amount, $\sim 50$ V, to improve the focusing of the primary electrons while they are still slow). The grid $G_2$ is negatively biased relative the specimen and thus prevents the electrons which have lost energy in the interaction with the specimen (i.e. those whose interaction has been inelastic) from reaching the screen.

Since low-energy electrons penetrate only to a depth of a few atomic layers, in LEED the surface is extremely susceptible to contamination resulting, for example, from adsorption of residual gases. Owing to this fact it is necessary to clean the surface carefully and to work in an ultra-high vacuum $(10^{-9} - 10^{-10}$ torr$)$. Under such conditions, the surface remains clean for $10 - 100$ min, enough time for the measurement to be carried out. Another experimental difficulty concerns the focusing of very slow electrons. The space charge does not allow focusing of the beam to a diameter less than 0.5 mm at currents of an order of 1 μA. To obtain a better focus, special arrangements have to be employed.

If the specimen is a poor conductor, it becomes charged by low-energy electrons. The effect is eliminated by periodically modifying the energy of the impinging beam (e.g. at a frequency of 50 Hz) from small to higher values. At elevated energies, the secondary emission coefficient for the

*Fig. 92:* The apparatus for LEED and Auger spectroscopy (courtesy of Varian Corp.).

specimen is usually greater than unity and so the charge accumulated during the period of impingement by low-energy electrons is liberated during the subsequent period with the high-energy electrons.

The resultant diffraction pattern is a function of the arrangement of the first few surface atomic planes of the specimen. The condition for a diffraction maximum is again equivalent to the Bragg condition, the lattice, however, having more the character of a two-dimensional structure. If the specimen is polycrystalline, i.e. if all possible crystal directions are present, the diffraction pattern will not yield any discernible spots. Therefore the method is only applicable for monocrystalline surfaces or for surfaces with some significant orientation. This enables contamination of the surface by layers of atomic thickness to be detected.

The theory of LEED is not yet sufficiently well elaborated to provide the structural factors determining the intensity of a particular maximum.

Experimental investigations using LEED have developed recently to the point where commercial apparatus for the technique has appeared on the market. Such an apparatus is shown in Fig. 92.

Fig. 93 shows an example of the diffraction pattern of the surface of pure nickel. On examination of the surface of some substances it has been found that the surface structure often has interesting properties whose

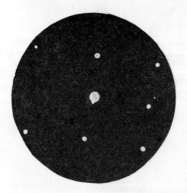

*Fig. 93:* Diffraction pattern of pure Ni (110) surface obtained by LEED (76 eV).

existence was not anticipated. It was found, for example, that the surface structure of substances with covalent bonding is frequently fundamentally different from that of the bulk. Subsequent research seems to indicate that special superstructures also occur in substances with other types of bonding.

# 5.4 X-ray Methods

### 5.4.1 X-ray Diffraction

The theory presented in Sect. 5.3.1 was originally developed for X-ray diffraction in crystal lattices and only later was it employed for electron diffraction.

In thin film physics electron diffraction is used more frequently than X-ray diffraction, but the latter is used where accurate results are required. One difference between the two methods is in the deeper penetration depth of X-rays, with the consequence that the diffraction spots are about $10^3$ times weaker compared with those of electrons. The greater penetration is also the reason why X-ray diffraction is more suitable for thicker specimens. But diffraction patterns may be obtained even for very thin specimens (e.g. down to 5 nm). By using an X-ray tube with a very sharp focus, the intensity may be increased and thus the exposure time reduced to about one-fourth. On the other hand, diffraction angles are much greater for X-rays, enabling lattice parameters to be determined with much higher precision than with the help of electrons. The diffraction pattern is produced by the whole thickness of the film, or possibly even by the substrate, which is in some cases advantageous for the determination of lattice constants.

X-ray diffraction may be also employed for the study of mechanical stresses in films (see Sect. 6.1.1).

From the width of diffraction lines it is possible to infer the size $a$ of crystallites in polycrystalline films according to the following expression:

$$a = \frac{\lambda}{D \cos \vartheta} \tag{5.9}$$

where $D$ is the angular width of the diffraction line at the half-maximum of its intensity, $\lambda$ is the wavelength of the radiation used and $\vartheta$ is the Bragg angle. In this way the dimensions of crystallites in the range of 5 to 120 nm may be determined. Although similar measurements may also be carried out with electron diffraction, this method is only applicable to crystals smaller than about 10 nm owing to the much shorter wavelengths of electrons.

### 5.4.2 X-ray Microscopy

X-ray microscopy is an appropriate technique for making crystal lattice defects visible. A narrowly focused beam of X-rays penetrates a specimen and after undergoing partial diffraction it reaches a photographic plate. The specimen and plate are parallel and they are moved so that the radiation successively probes different places on the specimen (in a way similar to that used in the scanning microscope). Defects in the whole depth of the specimen are visibly displayed on the screen and the area examined may be large.

The resolving power is of the order of 10 nm. The method is well suited to the observation of dislocations and has been used for the study of epitaxial semiconducting films (e.g. of germanium, silicon and GaAs).

## 5.5 Auger Spectroscopy

This method, which has recently been undergoing rapid development, is in fact a kind of supplementary technique to the LEED method. We have seen in Sect. 5.3.2 that for the formation of diffraction patterns only those electrons are used which have not suffered loss of energy. Inelastically scattered electrons are filtered off by the decelerating potential of the grid. These electrons can, however, provide valuable information because the magnitude of the energy loss is characteristic for a given surface.

There are two possible types of energy loss: ionization loss and the loss effected by the Auger mechanism. In the former case, a primary electron with energy $E_p$ excites another electron to which it transfers the ionization energy $E_i$ and returns to the vacuum as an inelastically scattered electron of energy $E_p - E_i$. The quanta $E_i$ are characteristic for a given substance because they are determined by its spectrum of energy levels. Thus when the energy distribution of the electrons is measured, the maxima corres-

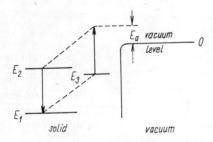

Fig. 94: Energy-level diagram illustrating the Auge° effect.

ponding to the inelastically scattered electrons occur at a certain distance from the maxima of elastically scattered electrons, and they shift in accordance with a change in the maximum of the primary energy.

In the second case the process is more involved: A primary electron excites an electron in the substance from an energy level $E_1$ (Fig. 94). An electron from a higher level $E_2$ makes a transition to the lower empty level, while the energy difference may be either liberated in the form of a photon (which is probable at higher energies, i.e. above 500 eV) or transferred to another electron, e.g. to that in the level $E_3$. This electron may be then emitted as an Auger electron with energy $E_a = E_1 - E_2 - E_3$. All three levels involved are characteristic for a given substance and so are the energies of the Auger electrons.

The levels are not sharp (they are more like bands), so the natural width of Auger spectral lines is relatively large (up to 10 eV). The lines are denoted by symbols common in X-ray spectroscopy (shells $K$, $L$, $M$, etc.; thus, e.g., $KLL$ indicates that the energy liberated by the transfer between $L$ and K has been transferred to an electron in the $L$ shell).

The depth from which the Auger electrons arise depends on their energy. Experiments reveal that for an energy of the order of 100 eV, the depth amounts to several atomic layers. The Auger effect therefore gives information of the composition of the surface layer of a substance.

To obtain the spectrum of Auger electrons, very accurate measurements must be made of the energy spectrum of the secondary electrons emitted by the surface, when exposed to a bombardment by electrons with energy of the order of hundreds of eV. For this purpose, the same system may be used as that in LEED. In the present case, we detect, however, electrons with lower energy than that of the primary electrons. The retarding curve is measured, i.e. the dependence of the collector current on the retarding potential between the specimen and the collector. At a given retarding potential the current consists only of those electrons that have energy sufficient to surmount the retarding field. By differentiating the retarding curve, we obtain the energy distribution of electrons $N(E)$, i.e. the relative number of electrons with energy $E$. The maxima corresponding to Auger electrons are mostly very small. To make them more pronounced, the curve of the energy distribution is again differentiated and salient double maxima turn up in the positions of the Auger maxima (see Fig. 95). Both differentiations are effected by means of electronic differentiation circuits.

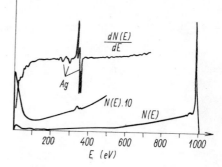

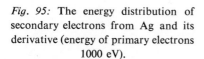

Fig. 95: The energy distribution of secondary electrons from Ag and its derivative (energy of primary electrons 1000 eV).

The identification of elements on a surface is facilitated by the fact that several lines belong to each of them. (Only hydrogen and helium cannot be detected by this method as they do not exhibit Auger transitions.) The positions and shapes of the lines are influenced by chemical bonding, which may be identified in some cases.

The smallest detectable amount of material depends on the actual experimental conditions. In the optimum case it is about $10^{-3}$ of a monolayer.

An Auger spectrometer in a cylindrical arrangement has also been developed enabling oscillographic recording of the spectrum to be made and thus allowing observation of rapid changes in surface composition.

# CHAPTER 6

# PROPERTIES OF THIN FILMS

## 6.1 Mechanical Properties

The mechanical properties of thin films play a very important role in every application because the stability of thin film systems depends on them. An internal stress in the film and insufficient adhesion to the substrate may lead to cracks in the film or to its peeling from the substrate. This explains why increasing attention has been paid to the problem recently.

The mechanical properties are, as are all other properties, largely determined by the film structure and that, in turn, is determined by the method of deposition. As has been shown in the earlier chapters, there are many factors influencing the structure and they are often difficult to check. A considerable difficulty arises when the results obtained are to be interpreted and unambiguous relations determined which would relate individual qualities characterizing mechanical properties to the parameters of film preparation. Difficulties have also been met in the experimental examination of mechanical properties. From the theoretical point of view it would, of course, be much simpler to work with monocrystalline epitaxial films. Owing to experimental problems, however, most results have been obtained on polycrystalline films, which render interpretation difficult. Recently however, studies have been carried out on epitaxial films.

The first papers on mechanical properties dealt in general with metals and some of the dielectrics. General theoretical concepts were suggested to explain the results obtained. Although there are theoretical models for the explanation of basic mechanical properties, satisfactory agreement between theory and experiment is still far from being achieved. The greatest attention has been given to the study of stress-strain curves, to the occurrence of plastic deformations and creep, to tensile strength and to the origin of internal stress.

## 6.1.1 Experimental Methods for Measurement
## of Mechanical Properties of Thin Films

($a$) The most frequently used method for internal stress measurement in films is the static method. In this method the film is deposited on a thin substrate in the form of rectangular strip or square plate. The substrate is fixed either at one end (cantilever beam) or is supported at both ends by blades. The stress in the film causes a strain in the substrate and the displacement of the free end is measured in the first case, that of center is the second case. The displacement is measured by interferometric techniques, electromechanical techniques (one of them is analogous to the stylus method see Sect. 3.5.1) or those using capacitance.

The stress is calculated from the displacement $\delta$ under the assumption that adhesion of the film to the substrate is strong, that no plastic deformation or creep occur on the interface and that the deflection is determined purely by the mechanical properties of the substrate, i.e. the film thickness is negligible compared to that of the substrate. The stress $S$ is then given as

$$S = \frac{\delta E_s D^2}{3L^2 t(1 - \varkappa_s)} \left(1 - \frac{E_f t}{E_s D}\right) \qquad (6.1)$$

where $E_s$ and $E_f$ are Young moduli for the substrate and film, respectively, $\delta$ the displacement, $t$ the film thickness, $D$ the substrate thickness, $L$ the length of the substrate and $\varkappa_s$ Poisson's ratio for the substrate. Since $t \ll D$, the relation may be rewritten as

$$S = \frac{E_s D}{6rt(1 - \varkappa_s)} \qquad (6.2)$$

where $r$ is the radius of curvature of the deflected substrate.

Substrates are usually mica or glass plates about 0.1 mm thick. The stress in a film may also be measured by the dynamic method. A thin (several $\mu$m) glass membrane coated by metal on one face is stretched on a metal ring and placed in a vacuum apparatus. After exhaustion it is mechanically agitated. The fundamental resonance frequency is usually about 1 to 2 kHz. During the film deposition on the surface of the membrane, modification of the resonance frequency occurs partly due to changed mass and partly due to a stress effect. The change in the mass may be monitored by simultaneous film deposition on a vibrating quartz crystal and the stress may then be calculated from the relevant theoretical equations. The exactness of measurement depends on the accuracy of the measurement of frequency, membrane thickness and the specific masses of the glass and film. The reproducibility of the membrane attachment to the ring is also of importance (for more further details see [6]).

Stress in films may also be determined by using the fact that it produces widening of diffraction fringes or spots in the diffraction of electrons or X-rays. Better results are obtained from X-ray diffraction because in this method the Bragg angles are greater.

For stress parallel to a film surface, the following relation holds:

$$S = \frac{E_f}{1 - \varkappa_f} \frac{d_0 - d}{d_0} \tag{6.3}$$

and for the perpendicular stress

$$S = \frac{E_f}{2\varkappa_f} \cdot \frac{d - d_0}{d_0} \tag{6.4}$$

where $d_0$ is the lattice constant of the bulk material without stress and $d$ is the lattice constant measured in the film under the measured stress, and $\varkappa_f$ is Poisson's ratio for the film. The deficiency of the method lies in the difficult separation of the line broadening caused by the internal stress from the analogous effect resulting from the small size of micro-crystallites (see Sect. 5.4.1).

In principle, other properties which are known to depend on the stress may be utilized for stress measurement. They include the occurrence of anisotropy in ferromagnetic films, changes of forbidden band width of semiconductors and modification of electrical resistance. The minimum measurable forces acting on a unit area of a film are different with different techniques and depend on the manner of recording. They vary from units down to several thousandths of $N/cm^2$.

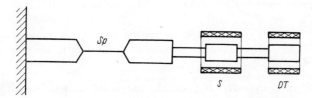

*Fig. 96:* Apparatus for stress-strain measurement. $Sp$ — thin film (specimen); $S$ — solenoid; $DT$ — differential transformer.

(b) The measurement of film strain under the load of a given stress together with the measurement of the tensile strength may be carried out with the system shown schematically in Fig. 96. A specimen $Sp$ is fixed between two jaws, one of which is fixed and the other connected with a mag-

154

net inserted through a solenoid $S$ (or with another type of device providing a defined force). Upon passage of current through the solenoid, tension is applied to the film and its strain is recorded either by a differential transformer $(DT)$ or by some other technique, e.g. optically. This method, which was originally developed for measurements on whiskers (thin needle-like monocrystals), is used mainly for films thicker than 0.1 μm.

Another method is based on mounting the film under examination at one end of a hollow cylindrical tube in which excess pressure is established; this leads to bulging of the film so that it forms a kind of spherical cap. The deformation may be observed interferometrically. The well-known Newton interference fringes arise and from their departure from a circular shape the film anisotropy may be simultaneously evaluated.

When using the pressure difference $p$, the relevant equation is

$$p = \frac{2t}{r}\left[T_0 + \frac{E_f a^2}{3p^2(1 - \varkappa_f)}\right] \tag{6.5}$$

where $r$ is the radius of the cap, $a$ the radius of the film, and $T_0$ the stress in the film without the action of the pressure difference (the other symbols have the same meaning as in the preceding equations). The expression in brackets gives the total stress in the film.

The bulge is not so much a spherical cap as a circular paraboloid or hyperboloid, in which case it is necessary to determine its fundamental parameters in a microscope and use a more complex formula. A difficulty with this method is insuring reliable attachment of the film to the rim. The best way to do this is to deposit the film on the actual place on which it is to be measured. This has been done with gold films which have been

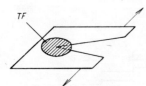

TF

Fig. 97: Schematic drawing illustrating a simple measurement of thin film tensile strength.

evaporated on a cylinder of rocksalt, the inner part of which was then dissolved by a water jet so that a self-supporting circular film was produced fixed to the rim.

To measure mechanical strength, a method has been devised in which the film is deposited on the jacket of a cylinder, the film adhesion being weak or its magnitude known. The cylinder is then rotated and at a certain speed the film cracks because of the effect of centrifugal forces, which may easily be calculated.

Another strength-measurement method uses the configuration shown in Fig. 97. The film is deposited on a slotted plate and forces are applied along the directions shown by the arrows. The technique affords the possibility of observing directly the process of film destruction in an electron microscope. Quantitative results are, however, only approximate owing to the complexity of the configuration.

### 6.1.2 Stress in Thin Films

Stress in a thin film consists basically of two components: the thermal stress arising during variations of temperature due to the difference in the thermal expansion coefficients of the film and the substrate, and the 'intrinsic' one, which depends on a number of factors.

When films are prepared by means of evaporation on a hot substrate or thermal decomposition, differences in the contraction of the film and the substrate, which mainly appear during the cooling of the system, generate a thermal stress given by

$$S = (\alpha_f - \alpha_s) \, \Delta T \, E_f \tag{6.6}$$

where $\alpha_f$, $\alpha_s$ are the average coefficients of expansion for the film and substrate, $\Delta T$ is the temperature of the substrate during the film deposition minus the temperature at measurement. The value of $\alpha$ for glass is $\sim 8 \cdot 10^{-6}/°C$, 10 to $20 \cdot 10^{-6}/°C$ for various metals and 30 to $40 \cdot 10^{-6}/°C$ for the alkali halides. If $\Delta T$ is positive, a tensile stress arises; if negative, a compressive stress is found.

The thermal stress is significant whenever the intrinsic stress is small. In films of low melting-point metals, such as In, Sn or Pb, intrinsic stress disappears rather rapidly at room temperature and reaches values smaller than $5 \cdot 10^2 \, N/cm^2$. The ratio between intrinsic and thermal stress depends, of course, on the actual method of preparation. For example, according to Chopra, with nickel films deposited on soft glass at a temperature of 75 °C and measured at 25 °C, the thermal stress is about 5% of the total stress. At higher temperatures thermal stress is dominant.

When discussing the role of thermal stress, we have to keep in mind that the actual temperature at which the film grows may differ considerably from the substrate temperature measured by a thermocouple. The actual temperature of condensation sites during evaporation is determined also by thermal radiation of the evaporation source and by latent heat liberated during the condensation. The former factor is further dependent on the reflectivity coefficient of the surface and that may vary during the film growth. Thus the actual surface temperature is usually difficult to determine.

Intrinsic stress may result from a number of causes and may reach large values for metals of the order of $10^4 - 10^5$ N/cm². The stress may have a tensile or compressive character; for metals and dielectrics deposited at room temperature it is mostly tensile. The character of stress may also vary with depth and this is the reason for curling of a film after it is peeled from the substrate. It seems that the average stress is almost independent of film thickness. Apparently it does not depend strongly on the substrate material but on the substrate temperature, deposition rate, angle of incidence, etc. Tempering usually reduces intrinsic stress and a certain temperature exists at which stress is minimal.

One of the causes of intrinsic stress may be that the actual temperature of a film may differ considerably from that of the substrate during deposition and may also vary throughout the whole process, so that the film grows successively under different conditions. Another effect contributing to stress is phase transitions (change in crystalline modification, stoichiometry) which can take place during deposition and which are accompanied by volume changes.

In thin films less than 10 nm thick, a difference in the surface tensions at the interface with substrate $(\sigma_1)$ and at that with air $(\sigma_2)$ may also lead to a stress which is then given as

$$P = \frac{\sigma_1 - \sigma_2}{t} \tag{6.7}$$

where $t$ is the film thickness. For typical values $\sigma \approx 10^{-4}$ J/cm² and at the thickness mentioned above the stress is of the order of $10^4$ N/cm².

Electrostatic effects also have consequences. The presence of free electric charges on crystallites increases their free energy and may thus lead to a change in geometry (crystallites attract or repel each other) and thus modify the stress.

A difference in the lattice constants of film and substrate ('misfit') results in an accommodation of the first atomic planes of the film to the parameters of the substrate, the lattice is deformed somewhat, and dislocations arise, as has been noted in connection with epitaxial growth. Dislocations as well as other lattice defects, whether point defects or other ones, may also increase stress. An important role is played by the boundaries of crystallites during the coalescence of which stress increases.

The total stress in a film sometimes reaches values comparable to the film strength so that a collapse or fracture ensues depending on whether compressive or tensile stress is at work.

### 6.1.3 Mechanical Constants of Thin Films

The most important relationship for the determination of mechanical constants is that of the dependence of strain on applied stress. A typical graph of this relationship is illustrated in Fig. 98. The dependence is linear at first, then the slope decreases. A second measurement does not reproduce the initial curve but yields a displaced curve (right, steeper curve) and this

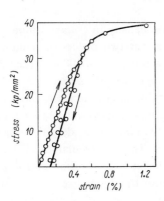

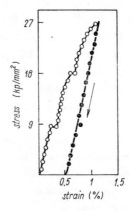

*Fig. 98:* Stress-strain curves for monocrystalline (111) Au film.

*Fig. 99:* Stress-strain curve of Au film with two regions exhibiting creep behavior (at stresses 9 and 18 kp/mm$^2$).

result is then reproduced with repeated measurements. This behavior shows that in no segment of the initial curve, i.e. not even at very small stress, is the strain completely elastic. If stress is applied to a film for some time (some minutes), creep may be observed, i.e. elongation of the film at constant stress. The effect is illustrated in Fig. 99, where stress has been kept constant for some time at values 9 and 10 kp/mm$^2$ respectively. If the elongation reaches $1-2\%$ a film fracture occurs which is preceded by a local plastic deformation. Nevertheless, at fracture the deformation is still three-quarters an elastic one. The elongation is much smaller than that normal for annealed bulk materials. The character of the deformation and the tensile strength are the same for polycrystalline and monocrystalline films. The elasticity moduli are in agreement with the bulk values.

High tensile strength is an important quality of thin films. The tensile stress needed for film fracture is higher than that for the typical annealed bulk material and still much higher than that of hard-drawn materials. The strength is sometimes a function of film thickness (e.g. Cu). Its value is in the range of $10^3$ to $10^6$ N/cm$^2$.

To explain high tensile strength, several hypotheses have been suggested. One of them ascribes responsibility to a layer of oxide on the surface which reduces the mobility of dislocations. It is obvious, however, that the real cause is to be found in some structural property connected with the method of thin film deposition by evaporation, since the high tensile strength is also observed in films of gold which have neither oxide nor any other layer on the surface, whereas this special feature is missing in the films prepared by abrasion of bulk material.

It has been found that the dislocation density in epitaxial films is in the range of $10^{10}$ to $10^{12}$/cm$^2$, which is about two orders of magnitude more than in bulk material. In films with a thickness $<100$ nm, the dislocation lines run mostly from one face to the other. Thus their movements may be impeded by their surface stabilization. In addition, the movement is obstructed by defects inside the film, which is perhaps the decisive factor. There are many kinds of defects in evaporated films, point defects being the most important according to some authors. The theory maintains that to stop dislocation movement completely, a defect concentration of $10^{10}$/cm$^2$ is adequate, which is quite a probable value for thin films. From stress-strain curves it appears that even at the greatest stresses the number for newly formed and transferred dislocations remains low, which is the condition of high tensile strength. From this aspect the properties of thin films are analogous to those of forged materials, in which a great concentration of defects is produced intentionally by mechanical working.

Another mechanical property investigated has been a microhardness measured on some thicker films (several μm), that is, hardness measured on a very small area and a correlation of this with tensile strength has been found.

### 6.1.4 Adhesion of Thin Films

The adhesion of films to a substrate is an important property in almost all applications. It depends on the nature and the strength of the binding forces between the substrate and film.

A qualitative estimate of adhesion is obtained by the Scotch tape test, in which the film is lifted off the substrate by adhesive tape. Methods employing various techniques of film abrasion are also used, their results being dependent on the hardness of the film. For a direct measurement, a force normal to the surface is needed and this may be provided by an ultracentrifuge or ultrasonic vibrations. The results obtained, however, are unsatisfactory.

The scratch method has turned out to be the best one available to date. A rounded chrome-steel point is moved across the film surface and the load applied to it is gradually increased. The critical load value at which the film is tripped from the substrate is the measure of adhesion. The measurement is done by observation in a microscope.

For this purpose, commercial hardness testers with tungsten carbide or diamond points ($r \sim 0.05$ mm) may also be employed. The loads needed for film stripping range from several grams for poorly adhering films to hundreds of grams.

Some authors maintain that the adhesion is caused by the van der Waals forces. A linear relation for several different metals evaporated on different alkali halides has been found between the van der Waals energy related to unit area and the shearing force applied in the scratch method. As the heat of condensation depends on the van der Waals forces, the former may also be considered as a kind of a measure of adhesion.

If a metal which forms oxides is deposited on glass, a transition region of oxides is established between the substrate and the film proper which greatly enhances the adhesion. In such cases the adhesion increases in the course of time due to continuing diffusion of oxygen from the atmosphere. The process depends on the film structure: Microcrystalline films transmit oxygen better and so facilitate the process.

It has already been mentioned that films prepared by cathode sputtering usually adhere better than evaporated ones. This is the consequence of the higher energy of the impinging particles, which create defects on the substrate surface and so enhance the binding energy, and, in addition, contribute to the formation of a transition region between the substrate and the sputtered material.

Impurities on the substrate may weaken or strengthen the adhesion depending on whether the binding energy is decreased or increased. Adhesion is usually enhanced when more nucleation centers are present.

Thus the origin of adhesion must lie in the van der Waals forces, or in chemical binding, as, for example, with oxide layers. Another possibility that has been considered is that the adhesion of metals to alkali halides is connected with the formation of an electrical double layer, the adhesion force being thus of electrostatic nature. This hypothesis has been elaborated theoretically, but is not yet generally accepted.

## 6.1.5 Rayleigh Surface Waves

It is a well known fact that volume acoustic waves may be generated in a solid. The waves, of frequency ranging from 100 Hz to several MHz, are used in various piezoelectric and magnetostrictive elements, mechanical filters, etc. The waves may be either longitudinal or transverse. The velocity of transverse waves is given as

$$C_{tr} = \sqrt{\frac{G}{\varrho}}$$

where $G$ is the modulus of elasticity in shear and $\varrho$ is the density of the solid. Besides these waves, a plane surface of a body exhibits surface acoustic waves which were, first described by Rayleigh (in 1887) and are called Rayleigh surface waves and denoted by RSW or ASW (acoustic surface waves). The waves are slower than those of the bulk, their velocity being

$$C_R = C_{tr} \frac{0.07 + 1.12\,\varkappa}{1 + \varkappa}$$

where $\varkappa$ is the Poisson constant. The velocity is independent of frequency over a very broad range and has typical values of $2 \cdot 10^5$ to $6 \cdot 10^5 \, \text{cm} \cdot \text{s}^{-1}$. Thus the wavelengths vary for frequencies from 10 MHz to 10 GHz in the range from 0.6 mm to 0.2 µm. The damping in the direction into the bulk is exponential, but is very small in the direction of propagation. The dependence of the vibrational amplitude on the depth $d$ below the surface, mea-

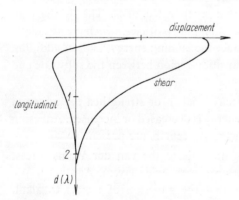

Fig. 100: The dependence of strain on depth below the surface for longitudinal and shear modes of vibration.

sured in units of wavelength of the given vibration, is shown in Fig. 100 for the longitudinal and shear mode, respectively. The decrease in amplitude with depth depends on frequency so that high frequencies are damped more strongly than low ones.

If there is a film of some other solid on a surface with a thickness equal to a fraction of a wavelength, then the wave becomes dispersive. This enables an acoustic waveguide to be constructed together with a number of other elements to which we shall return in the chapter on applications.

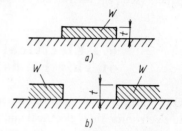

Fig. 101: Waveguide: (a) layer; (b) slotted.

A waveguide may be either of layer type (Fig. 101a) in which the film serves as the waveguide, or slotted in which the operative part is a slot between two films. In the former case the dispersion (i.e. variation of propagation velocity with frequency) is negative; in the latter it is positive.

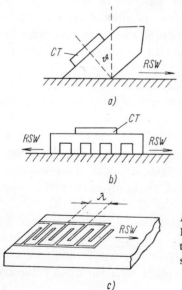

Fig. 102: Various modes of excitation of Rayleigh surface waves: (a) compression transducer (CT) and wedge; (b) compression transducer and metal comb; (c) interdigital transducer.

The frequency range depends on the film thickness. It has been found that as a broad-band element with a small dispersion the slotted waveguide is the more effective one.

The excitation of surface acoustic waves is brought about by the mechanical contact of an electro-acoustic transducer with the film via a wedge element (Fig. 102a), a compressively excited metal comb (*b*), or by an interdigital transducer (c). The latter methods may also be used for pick-up of vibrations. There are a number of other techniques for the excitation and pick-up of RSW (e.g. piezoelectrical, magnetokinetic, that using a PN junction in a semiconductor, etc.). The reader is referred to original papers (see [8]) for further references.

## 6.2 Electrical and Magnetic Properties of Thin Films

Owing to the striking development of applications of thin films of all kinds in electronics, their electrical and magnetic properties have become the focus of interest and it is already difficult to keep under review the large number of works concerned with them.

Modern solid state theory provides information on the energy structure of a given substance, i.e. the positions of electron energy levels and their occupation. The theory is known as the band theory since one of its most important conclusions is that electron energies in crystalline materials are arranged so as to form bands of allowed energies separated by forbidden bands. It is the configuration and occupancy of these bands which determines whether a given material is a metal, semiconductor or insulator.

The original band theory was developed for a fictitious infinite crystal. But in bulk crystalline material the surface (which in fact represents a large perturbation of crystal potential) modifies the density of the energy states within the vicinity of the surface and thus effects resulting from changes in the energy structure may be observed experimentally. Prominent among these effects is the occurrence of surface states in semiconductors, i.e. of allowed discrete energy levels which exist only at the surface and are forbidden inside the bulk.

If we pass from a crystal with one surface to a thin film, i.e. to a system with two surfaces, the changes in the energy structure are still more pronounced and so-called quantum size effects appear. In addition, mutual interaction of both surfaces will begin to play a role, which will influence the positions of the surface levels.

More detailed discussions of these effects would go beyond the scope of this book as it necessitates knowledge of quantum mechanics. It is intended here only to point out these circumstances, since they can play a great role in interpreting some of the effects in thin films.

The structure of thin film is usually very different from the ideal one assumed in theory.

Since the electrical and magnetic properties mainly depend on film structure, fundamental disagreements are often found between individual papers. This is because until quite recently the methodology of monitoring of film parameters during preparation was insufficiently advanced. Even now, when the requisite techniques are well elaborated, complete reproducibility is not always certain because monitoring of all the parameters is an arduous task, demanding time and specialized equipment, and far from all thin film laboratories are furnished with a complete set of measurement devices. In addition, it sometimes happens that the film properties are more sensitive to some factor than the measuring device itself (e.g. to a particular component of residual gas). We have to bear all this in mind when reviewing the results of individual papers.

The use of thin films as resistors has led to extensive study of conductivity, its temperature dependence, the effect of thermal processing stability and so on. The use of dieletric thin films in capacitors has necessitated the study of dielectrical properties, and the development of thin film transistors has led to detailed investigation of all electrical transport phenomena, etc.

### 6.2.1 Conductivity of Continuous Metal Films

The resistance of a metal film placed between two contacts may be obtained from the well-known equation

$$R = \varrho \, \frac{l}{S} \tag{6.8}$$

where $S$ is the cross section of the film, $l$ is its length and $\varrho$ is the specific resistivity. The question now arises whether this resistivity is equal to that of the bulk material. Measurements reveal that thin film resistivity is usually higher.

When we start from the basic concept that metal materials consist of a lattice built up from ions and electron gas, we come to the conclusion that resistance is caused by collisions of the free electrons with lattice defects. The longer the mean free path of electrons the higher the conductivity of the bulk material. An elementary theory yields the expression

$$\sigma_{\mathrm{B}} = \frac{Ne^2 \lambda_0}{mu} \tag{6.9}$$

where $N$ is the concentration of the free electrons, $\lambda_0$ the mean free path of the electrons in the bulk material, $m$ and $e$ the mass and charge of an electron, and $u$ the mean thermal velocity. An increase in the number of electron collisions results in a decrease in the conductivity, i.e. an increase in resistivity. The theory has further established that the collisions of electrons with the lattice do not lead to a decrease in the conductivity provided the lattice is perfect and that resistivity only arises because of collisions with imperfect lattices. The imperfection may be a real irregularity in the lattice structure (impurity, symmetry defect) or it may be a result of thermal displacement of a given lattice point.

According to the Matthiesen rule, resistivity may be divided into components corresponding to particular types of scattering. For a bulk material we may write

$$\varrho_B = \varrho_{ph} + \varrho_g \tag{6.10}$$

where $\varrho_{ph}$ relates to vibrations of the crystal lattice (phonon interaction) and $\varrho_g$ corresponds to the scattering at geometrical structure defects. The phonon interaction decreases with decreasing temperature and at very low temperature only the residual resistivity, corresponding to the interactions with defects remains.

Now, let us consider what happens if we reduce one dimension of a given sample. If the dimension remains much longer than the mean free path of electrons in the bulk material, no effect will be observed. However,

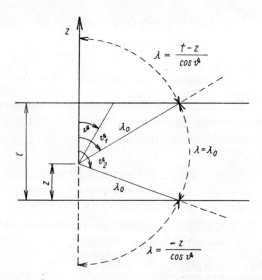

Fig. 103: For the discussion of the size effect in thin film conductivity.

as soon as the sample thickness is comparable with the mean free path, a new mechanism of scattering appears, namely, scattering on both surfaces of the film. A fraction of the electrons will reach the surface before they run the whole length of the free path, the effective length of the path will thus be smaller, which corresponds to an increase in resistivity. The resistivity may be now written as

$$\varrho = \varrho_{ph} + \varrho_g + \varrho_s \qquad (6.11)$$

where $\varrho_s$ is the resistivity component corresponding to the surface scattering. The scattering may be either diffuse, which means that electrons do not have any preferred directions after the interaction, or specular.

The first theory for completely diffuse scattering was elaborated by Thomson on the basis of geometrical considerations.

In Fig. 103 a film of thickness $t$ is schematically illustrated. Electrons are ejected at various angles from a certain point. The electrons may be divided into three groups: those moving at an angle $0 < \vartheta < \vartheta_1$, which arrive at the surface before they travel a mean free path $\lambda_0$; those with $\vartheta_1 < \vartheta < \vartheta_2$, which travel a mean free path; and those with $\vartheta_2 < \vartheta < \pi$, which again reach the surface prematurely. From this model Thomson concluded that the mean free path would be

$$\lambda = \frac{t}{2} \left( \frac{3}{2} + \ln \frac{\lambda_0}{t} \right) \qquad (6.12)$$

The conductivity is therefore

$$\sigma = \frac{te^2 N}{2mu} \left( \frac{3}{2} + \ln \frac{\lambda_0}{t} \right) \qquad (6.13)$$

The theory predicts that the conductivity of a thin film will be equal to 75% of that of the bulk material if the film thickness is just equal to the mean free path $\lambda_0$. The theory has, however, serious faults. To give an example: the value of $\sigma$ does not approach $\sigma_0$ in (6.13) for $t/\lambda_0 \to \infty$.

A more sophisticated analysis of the problem has been carried out by Fuchs, Sondheimer et al., who proceeded from the solution of the Boltzmann transport equation. This equation* describes the effect of external forces and collisional processes on the electron distribution function $f$ in its dependences on the time $t^*$ and co-ordinates $r$. The equation has the form

$$\frac{\partial f}{\partial t^*} + \vec{v} \cdot \mathrm{grad}_r f + \frac{\mathrm{d}\vec{v}}{\mathrm{d}t^*} \, \mathrm{grad}_v f = \left( \frac{\partial f}{\partial t^*} \right)_{col} \qquad (6.14)$$

where $v$ is the velocity of electrons and $(\partial f/\partial t^*)_{col}$ is the collision term. In the case considered the equation assumes a much simpler form. Since we are

---

* There is no place in this book to expound the problem in detail. The reader who is not acquainted with the Boltzmann transport equation may skip the following lines and continue from equation (6.21).

considering an equilibrium state, when $f$ does not vary with time, the term $(\partial f/\partial t^*)$ equals zero. External force is represented by an applied electric field, $F$. We assume that the force $F$ acts in direction $X$. As we also assume that the film is infinite in the directions $x$ and $y$, the distribution function does not vary with these co-ordinates. Thus

$$\dot{x}\frac{\partial f}{\partial x} = \dot{y}\frac{\partial f}{\partial y} = 0$$

and (6.14) is reduced to

$$\frac{dz}{dt^*}\frac{\partial f}{\partial z} - \frac{eF}{m}\frac{\partial f}{\partial \dot{x}} = \left(\frac{\partial f}{\partial t^*}\right)_{col} \tag{6.15}$$

(the dot denotes a time differentiation).

Now we have to determine the collisional term, i.e. the change in the distribution function brought about by collisions of the electrons. The method is analogous to that used in other problems. A relaxation time $\tau$ is introduced which is defined as the time needed for the change of the distribution function to decrease to $(1/e)$th of its maximum magnitude after the cause of the change is eliminated. If we denote the value of the unperturbed function as $f_0$ we may write

$$\left(\frac{\partial f}{\partial t^*}\right)_{col} = -\frac{f(t^*) - f_0}{\tau} \tag{6.16}$$

and hence the Boltzmann equation is now given by

$$\frac{dz}{dt^*}\frac{\partial f}{\partial z} - \frac{eF}{m}\frac{\partial f}{\partial \dot{x}} = -\frac{f(t^*) - f_0}{\tau} \tag{6.17}$$

The function $f$ is usually written in the form

$$f = f_0 + f_1(v, z) \tag{6.18}$$

where $f_1$ is the change of distribution function dependent on the velocity and position with respect to the boundaries of the film. The solution of the differential equation with imposed boundary conditions (i.e. of diffuse scattering on both surfaces of the film, $z = 0$ and $z = t$) yields

$$f_1 = \frac{eF\tau}{m}\frac{\partial f_0}{\partial \dot{x}}\left[1 - a\exp\left(-\frac{z}{\tau \dot{z}}\right)\right] \tag{6.19}$$

where $a = 1$ for the electrons with the velocity in the positive $z-$direction and $a = \exp\left(t/\tau\dot{z}\right)$ for oppositely moving electrons.

Once the distribution function is known, it is possible to determine the current density and conductivity relative to the bulk material. The result is

$$\sigma = \sigma_0 \left\{ 1 - \frac{3\lambda_0}{2t} \int_0^{\pi/2} \left[ 1 - \exp\left( -\frac{t}{\lambda_0} \right) \cos \vartheta \right] \sin^3 \vartheta \cos \vartheta \, d\vartheta \right\} \quad (6.20)$$

For thick films $(t \gg \lambda_0)$, we obtain an approximation in the form

$$\sigma = \sigma_0 \left( 1 + \frac{3\lambda_0}{8t} \right)^{-1} \quad (6.21)$$

and for thin films $(t \ll \lambda_0)$ we may use

$$\sigma = \sigma_0 \frac{3t}{4\lambda_0} \left( \ln \frac{\lambda_0}{t} + 0.4228 \right) \quad (6.22)$$

As we have seen, the boundary conditions were for the case of diffuse scattering. The theory may, however, allow for some fraction $p$ of the electrons being reflected specularly. The equations (6.21) and (6.22) then take the form

$$\sigma = \sigma_0 \cdot \left( 1 + \frac{3(1 - p) \lambda_0}{8t} \right)^{-1} \quad (6.23)$$

$$\sigma = \sigma_0 \frac{3t}{4\lambda_0} (1 + 2p) \left( \ln \frac{\lambda_0}{t} + 0.4228 \right) \quad (6.24)$$

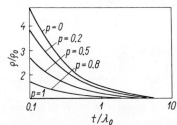

Fig. 104: $\varrho/\varrho_0$ as a function of $t/\lambda_0$.

The ratio of the number of electrons reflected specularly and diffusively depends on the nature of the interface. It may be expected that the film-substrate interface will have a different effect than the interface with air (or vacuum). Some account of this may be taken by introducing two different coefficients $p$ and $q$; then the approximate equation for thick films assumes the form

$$\sigma = \sigma_0 \left[ 1 - \frac{3\lambda_0}{8t} \left( 1 - \frac{p + q}{2} \right) \right] \quad (6.25a)$$

and for very thin films

$$\sigma = \sigma_0 \frac{3t}{4\lambda_0} \frac{(1 + p)(1 + q)}{1 - pq} \ln \frac{\lambda_0}{t} \qquad (6.25b)$$

Theoretical curves of $\varrho/\varrho_0$ versus $t/\lambda_0$ for different values of $p$ are given in Fig. 104.

Another electrical parameter of importance is the temperature coefficient of resistivity. It may be also expected that in general this coefficient will be different for a thin film and the bulk material. The coefficient is defined for bulk material (subscript B) and for thin film (subscript $f$) by the relations

$$\alpha_0 = \frac{1}{\varrho_B} \frac{d\varrho_B}{dT} \qquad \alpha_f = \frac{1}{\varrho_f} \frac{d\varrho_f}{dT} \qquad (6.26)$$

As noted already, the temperature-dependent component of resistivity originates from the interaction of electrons with phonons, whereas their scattering on lattice defects is practically temperature independent. The resistivity connected with the scattering on a surface is a function of the ratio of the thickness and electron mean free path, $\lambda_0$. Since the mean free path decreases with increasing temperature the resistivity produced by the surface scattering will in general vary with temperature.

Only for rather thick films is the surface scattering practically temperature independent and we may write in this case

$$\frac{d\varrho_f}{dT} = \frac{d\varrho_{ph}}{dT} = \frac{d\varrho_B}{dT} \qquad (6.27)$$

so that we obtain, by using (6.26),

$$\alpha_f \varrho_f = \alpha_B \varrho_B \qquad (6.28)$$

A comparison with experiment has revealed that for epitaxial films of gold the theoretical results agree well with the measured one for $t/\lambda_0 = = 4 - 5$ and $p = 0.8$. For polycrystalline films agreement has been obtained for $p = 0$. This is perhaps due to the fact that polycrystalline films have less smooth surfaces than monocrystalline ones.

The temperature coefficient of resistivity may be evaluated explicitly for very thin films $(t = 0.1\lambda_0)$, in which case

$$\alpha_f = \frac{\alpha_B}{\ln \dfrac{\lambda_0}{t} + 0.4228} \qquad (6.29)$$

A comparison of theoretical results with some experimental curves is shown in Fig. 105.

The results for thin film resistivity and its temperature dependence presented above may be used for the determination of the unknown parameters $\varrho$, $\lambda$, $p$. It has to be noted that, for example, the value $\varrho_B$ of resistivity

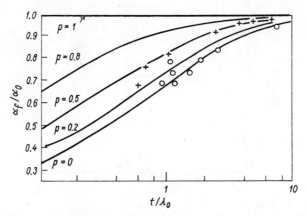

Fig. 105: $\alpha_f/\alpha_B$ as a function of $t/\lambda_0$; theoretical curves are solid; experimental points: +-Chopra and Bobb; $\circ$ — Leonard and Romey.

need not be equal to that of the bulk material. Thin films usually contain many more defects than the bulk material and thus the residual resistivity will usually be higher in the films. In ferromagnetic films, additional scattering centers may be represented by boundaries of magnetic domains.

The resistivity of a thin film may be modified by covering its surface with some material. The resistivity may increase or decrease depending on the film-overlay combination. An example of the variation in resistivity with the thickness of the surface coating is shown in Fig. 106.

The surface can also influence the resistivity in other ways. Surface gas adsorption, for example, can produce considerable change in the conductivity of evaporated and sputtered films, while impurities from solutions can affect the electrical properties of electrolytically deposited films. The induced surface charge is another possible cause of conductivity modification. This phenomenon, which is called the field effect, plays a much more prominent role in semiconductors than in metals, because it is in the former that substantial changes in occupation of surface states take place. The conditions under which the film has been prepared influence its properties in various ways. They determine crystal structure, type and concentration of defects, impurity content (e.g. of adsorbed gases), quality of both surfaces, etc. All these parameters have, as we have seen, a direct relation ot conductivity and its temperature dependence, which may therefore be modified over a very broad range by suitable choice of the preparation technique.

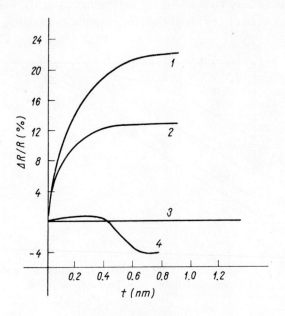

*Fig. 106:* $\Delta R/R$ as a function of coverage thickness: (1) 150 nm Ag, Ge coverage; (2) 22.3 nm Au, permalloy coverage; (3) 10 nm Al, Ge coverage; (4) 12.9 nm Cu, Cr coverage.

### 6.2.2 Conductivity of Discontinuous Metal Films

It has been found on studying the effect of thickness variation on thin film conductivity that at a certain thickness, which depends on a number of parameters, the conductivity of a film exhibits a sharp increase by several orders of magnitude (e.g. $10^6 \times$ multiple, see Fig. 107). The thickness corresponds to the stage at which the film becomes continuous and begins to show normal metal conductivity. Films of smaller thickness apparently consist of individual, electrically isolated islands, as is indeed confirmed by electron micrographs.

But even these films exhibit a certain conductivity which is not only much smaller in absolute magnitude than that of the bulk metal but shows different a temperature dependence as well; it usually increases very rapidly (exponentially) with temperature. This type of dependence supports the conclusion that a thermally activated process is involved in the conductivity of discontinuous films.

For an explanation of this conductivity two different concepts are invoked: $(a)$ charge transfer is due to the electrons thermally emitted from one island and captured by another island and $(b)$ transfer is effected by tunneling between two neighboring islands either via an air (vacuum) gap

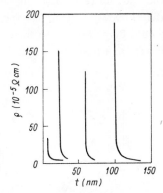

*Fig. 107:* The thickness dependence of the resistivity of Hg film evaporated onto glass at various substrate temperatures (from left to right: 20 K, 68 K, 78 K, 90 K, respectively) (Appleyard, Bristov).

or via the substrate. If the transfer is assisted by a strong electric field between the islands, the former case represents field-assisted thermionic emission, the Schottky effect. The actual field between the islands is much stronger than that calculated from macroscopic parameters (length of the film and applied voltage) since inside metal islands the field is practically zero and the whole potential fall is concentrated in the gaps. Calculations have established that in films of ordinary parameters at 300 K, the Schottky effect is dominant only in those which have relatively large inter-island spacings ($\geqq 10$ nm), whereas spacings of $2-5$ nm have led to dominance of the tunneling.

Thermionic emission between the islands is governed by a relation which is to a certain extent analogous to the Richardson equation for thermionic emission from a metal to a vacuum. Conductivity (to which the passing current is proportional) is given as

$$\sigma = \frac{BeT}{k} d \exp\left(-\frac{\varphi - Ce^2/d}{kT}\right) \qquad (6.30)$$

where $\varphi$ is the bulk work function of the film material, $d$ the distance between islands, $C$ and $B$ are constants, $k$ is the Boltzmann constant, $T$ the absolute temperature, $e$ the electronic charge. The term $Ce^2/d$ represents the effect of the image forces, which can substantially reduce the effective work function

172

at small inter-island spacings. If there is also an applied electric field $F$ between the islands, the effective work function will be further reduced by the Schottky effect

$$\varphi_{ef} = \varphi - \frac{Ce^2}{d} - C'e\sqrt{eF} \tag{6.31}$$

The question still remains open whether thermionic emission operates via the vacuum (the equations hold in this case) or via a conduction band of the substrate which has the character of an insulator. The second way would be preferable to the first because the metal-dielectric barrier is much lower than that of the metal-vacuum interface.

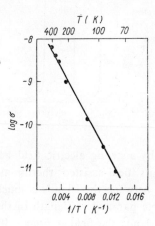

Fig. 108: Log $\sigma$ as a function of $1/T$ for Pt film with island structure (Neugebauer, Webb).

Thus, the conductance resulting from thermionic emission and the Schottky emission is characterized by a linear dependence of the logarithm of the current on $1/T$, linear characteristics at low fields ($\sigma$ is constant, as follows from (6.30)) and exponential dependence of the current on $\sqrt{F}$ with strong fields.

If the electrons travel via the substrate two cases may occur: if the distance between the islands is smaller than the mean free path of electrons in the substrate, the passage of current is effected by the mechanism mentioned; if, however, the distance is greater, the limiting factor is the conductivity of the substrate.

The tunneling of electrons between two conductors separated by a small dielectric gap has been investigated by a number of authors. The theory, using a quasiclassical approach, yields for the probability $D$ of the transmission of electrons through a barrier (transparency of barrier) of height $\Phi$ and width $d$:

$$D = \frac{\sqrt{(2m\Phi)}}{h^2 d} \exp\left(-\frac{4\pi d}{h}\sqrt{(2m\Phi)}\right) \tag{6.32}$$

where $h$ is Planck's constant. At a very small applied potential difference, the probability of transition from one particle of radius $r$ to another particle of the same kind is

$$P = Dr^2eV \tag{6.33}$$

The time required for an electron to travel the distance between two particles is inversely proportional to the transition probability so that the mobility (i.e. the velocity in an applied field of unit intensity) is

$$\mu = \frac{dP}{V/d} = Der^2d^2 \tag{6.34}$$

To induce a transfer of charge between two initially neutral particles, energy has to be supplied to overcome electrostatic coulomb forces. The energy amounts to $W = e^2/\varepsilon r$ where $\varepsilon$ is the dieletric constant. This is an activation energy which may be supplied, for example, in the form of heat. We speak here about activated tunneling. Charged particles are thus the 'charge carriers'. In a state of thermodynamic equilibrium, their concentration is given as

$$n = \frac{n_0}{n_0r^3} \exp\left(-\frac{e^2}{\varepsilon rkT}\right) \tag{6.35}$$

where $n_0$ is the total number of particles which are supposed to form a linearly ordered array.

By using the appropriate expressions we obtain for the conductivity

$$\sigma = ne\mu = \frac{e^2d \sqrt{(2m\Phi)}}{rh^2} \exp\left(-\frac{e^2}{\varepsilon rkT} - \frac{4\pi d}{h} \sqrt{(2m\Phi)}\right) \tag{6.36}$$

Thus we can see that even in this case $\sigma$ depends exponentially on $1/T$ and that at small voltages the current obeys Ohm's law. The dependence of $\sigma$ on inter-island spacing and size of islands is also strong.

The activation energy $W$ which we have introduced is actually modified by still two further factors. In the first place the energy is lower than the value obtained above because the charge is not transferred to infinity but only along a distance $d$. Therefore the energy is smaller and is given by

$$W = \frac{e^2}{\varepsilon r} - \frac{e^2}{\varepsilon(d - r)} \tag{6.37}$$

Furthermore, in strong fields the height and width of the barrier are modified by the field in such a way that the effective activation energy is further reduced. The modifying effect of the field $F$ is expressed by the relation

$$W_{ef} = \frac{e^2}{\varepsilon r} - \frac{2e\sqrt{(eF)}}{\varepsilon} + reF \qquad (6.38)$$

The second term leads to an exponential dependence on $\sqrt{F}$ in current-voltage characteristic at fields of $10^3 - 10^4$ V/cm. The last term is negligible for small $r$ and intermediate fields. In powerful fields the probability of the tunneling is so great that the whole process is limited by the rate of thermal excitation of electrons and the current becomes saturated.

Bearing in mind that tunneling occurs between very small particles, another possible effect has been considered. We have to remember that the states of electrons are quantized in small particles, and that the tunneling is possible only between allowed states. The ground levels are very narrow and are not therefore expected to cross, especially when there is a potential drop established between neighboring islands. On the other hand, the first excited level is broader $(\sim 10^{-3}$ eV for a particle of 10 nm radius), which enables the levels to cross at a given interspacing up to fields of several hundreds of V/cm. This model has been contested by pointing out that the widths of excited levels are actually narrower and that particles are of a variety of sizes and shapes so that their energy spectra are different, rendering the tunneling between the bands improbable.

The hypothesis that the process in some way involves the excitation of particles is substantiated by some effects (luminescence and electron emission into the vacuum) accompanying the passage of current through extremely thin films. Although these effects have not been fully explained yet, they indicate the existence of excited particles. The surplus energy of the particles may either be radiated out or used to surmount the work-function barrier.

In addition to tunneling through a vacuum, tunneling through the film-substrate interface barrier is also possible. The probability of such tunneling is even greater because the barrier height is less (work function of the metal minus the electron affinity of the dielectric) and the activation energy is smaller (the reduction depends on the ratio of the dielectric constants). Besides direct tunneling, tunneling through traps and defects may also contribute to the current. This could also be the explanation of the quantitative disagreement between experiment and the theory of simple tunneling, which predicts much lower conductivity than is observed.

The great intensity of the observed current has also been explained by the fact that free carriers created by both thermionic emission and tunneling are injected into the dielectric material of the substrate, where they are transferred from trap to trap by a hopping process.

The temperature-dependence of currents passing through discontinuous films is influenced by many factors. In general the temperature change induces a change in the conductivity of both individual islands and substrate. Owing to thermal expansion, a change is to be expected in the dimensions and distances, which in turn may have a profound effect on the probability of tunneling. The thermionic energy supplied to the electrons is also changed. Furthermore, secondary changes may occur which can substantially modify the current. These are changes in sorption on the surface of the substrate and islands, as well as changes in the morphology of the film proper caused by tempering. An exhaustive interpretation of the temperature-dependence in every case is a very complicated task. Experiments have yielded both positive and negative temperature coefficients of conductivity over a broad range. An increase with the temperature is more frequent and indicates the dominant role of thermal activation in the process. On the other hand, in cases when the thermal expansion of the substrate greatly exceeds that of the film, an increase occurs in inter-island spacing, which leads to a decrease in conductivity (this is typical for films on Teflon, which has a very high coefficient of thermal expansion).

In concluding this section we shall deal briefly with the conductivity of films which form a transition group between films with crystallite structure and continuous ones. These are the films in which the individual islands have been electrically connected but which still contain a large number of gaps. The behavior of such films is determined by a combination of the properties of the metallic bridges and the gaps between them. Thus the conductivity is given by the sum of the metallic contribution and that of tunneling or thermionic emission over empty regions. An important role is played here by scattering on boundaries and by all the factors which can in some way affect the properties of the surface (adsorption, surface oxidation) and the structure of metallic bridges (changes brought about by tempering, ageing, etc.). The temperature-dependence of conductivity is therefore very complicated.

Similar properties are also exhibited by films of greater thickness which have a porous structure owing to the method of preparation.

### 6.2.3 Electrical Properties of Semiconducting Thin Films

The experimental investigation of semiconducting thin films has to deal with still greater difficulties than that of metallic films. This is because of the great dependence of conductivity and related properties on slight concentrations of impurities and defects in the crystal lattice. As a result the diversity in the results obtained by different authors on similar materials has been much greater than that for metals, and the interpretation of the data has proved to be very difficult. Although considerable advances have recently been achieved in this field, no generally valid rules can be presented for semiconductors and most papers deal only with properties of a selected material prepared under defined conditions. In addition to the technological problems, difficulties arise from the fact that the properties of semiconductors even in bulk form are much more complex and diverse than those of metals.

If we disregard all defects and consider a thin film of an ideal semiconductor with perfectly parallel boundary planes, we can, as with metals, investigate the effects of size, especially the relation between thickness and conductivity of a film, i.e. between thickness and mobility of free charge carriers.

The surface properties and effects are generally of great importance in semiconductors and they have a special influence on their electrical properties. This influence results from the fact that the field produced by surface charge penetrates to a great depth in the semiconductor and thus affects the elctrical transport phenomena (i.e. transport of charge carriers).

Only in special cases are we able to ignore this field, which causes a bending of energy bands at the surface. Such cases occur if we are dealing with a specimen of great thickness or with a film whose thickness is small compared to the penetration depth of the field, i.e. the Debye length

$$L_{\text{D}} = \left( \frac{4\pi\varepsilon k T}{e^2(n_0 + p_0)} \right)^{\frac{1}{2}} \tag{6.39}$$

where $n_0$ and $p_0$ are the concentrations of electrons and holes, respectively, in the bulk material and $\varepsilon$ the static dielectric constant.

We can again distinguish two types of scattering: the surface and the bulk one. Each of these interactions has its characteristic relaxation time and the resultant relaxation time $\tau$ is given as

$$\frac{1}{\tau_{\text{f}}} = \frac{1}{\tau_{\text{s}}} + \frac{1}{\tau_{\text{o}}} \tag{6.40}$$

We may obtain an estimate of $\tau_s$ by dividing the total thickness of the film, by the mean velocity $v_z$, that is

$$\tau_s = \frac{t}{v_z} = \frac{t}{\lambda}\,\tau_o = \gamma\tau_o \tag{6.41}$$

where $\lambda$ is the mean free path of the electron defined by

$$\lambda = \tau_o v_z\,, \quad \gamma = t/\lambda$$

The $\lambda$ is related to other parameters by

$$\lambda = \mu_0 \frac{h}{e}\left(\frac{3}{8\pi}\,n_0\right)^{\frac{1}{3}} \tag{6.42}$$

where $\mu_0$ is the mobility and $n_0$ the concentration of the free carriers in the bulk material of $n$ type. The typical values for a degenerate semiconductor are $n_0 = 10^{18}\,\mathrm{cm}^{-3}$, $\mu_0 = 1000\,\mathrm{cm^2/Vs}$, which give $\lambda \sim 20\,\mathrm{nm}$.

For the mobility in thin films, we may use an expression analogous to that for the bulk, i.e. $\mu_f = e\tau_f/m^*$, where $m^*$ is the effective mass of the carriers. By combining (6.40) and (6.41) we obtain

$$\mu_f = \frac{\mu_0}{1 + 1/\gamma} \tag{6.43}$$

Thus the mobility decreases with decreasing film thickness. If all the electrons are not scattered diffusely but a fraction, $p$, of them is reflected by specular scattering, equation (6.43) should be modified to

$$\mu_f = \frac{\mu}{1 + (1 - p)/\gamma} \tag{6.44}$$

In thick films $(t > L_D)$, we may not normally neglect the bending of energy bands caused by the surface charge and the size effects in conductivity should be ascribed mainly to this factor.

If we denote the surface potential drop by $V_s$, we can introduce a normalized quantity $v_s = eV_s/2kT$ and effective distance $L_{ef}$ from the surface to the center of the space charge:

$$L_{ef} = \frac{v_s}{F_s}\,L_D \tag{6.45}$$

where $F_s$ is the space-charge distribution function, which decreases rapidly with increasing $v_s$. If the bands are bent only slightly, i.e. $v_s$ is small, then $L_{ef}$ approaches $L_D$. When calculating the surface conductivity (of a non-degenerate semiconductor of $n$ type) we have to take into account that the electrons in the bending region are accelerated if the bands are bent down-

178

wards (accumulation surface layer), or retarded (depletion surface layer) in the case of upward bending and that only those electrons reach the surface which have sufficient energy to surmount the retarding barrier.

In the former case, theory gives the following expression for surface mobility:

$$\mu_s = \frac{\mu_0}{1 + \dfrac{1}{L_{ef}}(t - p)(1 + v_s)^{1/2}} \tag{6.46}$$

The corresponding expression for the depletion layer is more complex.

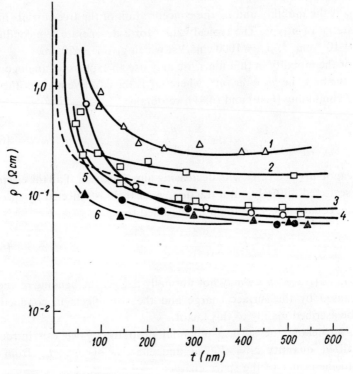

*Fig. 109:* The thickness dependence of Ge film resistivity for various evaporation rates: (1) 7.2 nm/min; (2) 200 nm/min; (3) 75 nm/min; (4) 470 nm/min; (5) 210 nm/min; (6) 120 nm/min; the dashed line is the theoretical curve for $p = 0$, $\lambda = 50$ nm.

For the case of flat energy bands, we may apply Fuchs's theory and obtain results similar to those obtained for metals. A comparison of experimental results with this theory for evaporated polycrystalline films of germanium is given in Fig. 109.

In addition to normal size effects, semiconductor and semimetal* films may exhibit, under certain conditions, so-called quantum.-size effects. They appear when the thickness is comparable not only with the mean free path but also with the effective wavelength of electrons because then the transverse component of electron momentum becomes quantized and, consequently, the energy spectrum of electrons become quasi-discrete. For this additional quantization to be observable, it is necessary that the energy spacings $\Delta E$ be larger than the broadening of the levels due to thermal and other miscellaneous scatterings. Conditions suitable from this point of view occur in materials with small effective mass ($\sim 0.01$ m), large mobility (of the order of thousands of $cm^2/Vs$) and thickness of about 100 nm. Owing to this effect, the bottom of the conduction and the top of the valence band are separated by an additional energy interval $\Delta E$. This mutual shift is especially evident in cases when the bands partly overlap as, for example, in Bi and Sb. There we may expect the most profound effect of thickness variation on electrical transport properties. The changes in the properties vary periodically with the thickness and their amplitudes increase with decreasing temperature. In films in which $\Delta E$ is larger than the original overlap the metallic character of the conductivity is replaced by the semi-conductive. On both bismuth and antimony such effects have indeed been observed.

The influence of surface charge on the conductivity of semiconductors may be utilized by modulating it with an external electric field. This field effect, already mentioned in connection with metals, is employed in the thin film transistor. We shall return to this subject in Chapter 7.

We should note that a large number of works has been devoted to thin films of semiconductor systems of II — VI types (CdS, CdSe, PbSe, PbTe, etc.) and III — V types (GaAs, GaP, InAs, InSb, GaSb, etc.) in view of their interesting properties and possible applications. These materials have either remarkable photoconductivity, photovoltaic effects, luminiscent properties, or they represent suitable materials for lasers and microwave generators. To discuss all their properties is beyond the scope of this book. Let us, however, note that these materials dissociate with comparative ease and hence their preparation in the form of thin films is in itself a very complex technological problem.

---

* for example Bi, Sb-metals, where the valence and conductivity band are only slightly overlapped.

### 6.2.4 Galvanomagnetic Effects in Thin Films

The most widely known galvanomagnetic effect is the Hall effect. If a conductor with current $I_x$ in the $x$-direction is placed in a magnetic field $(H_z)$, perpendicular to the current direction, then the charge carriers are subjected to a deflecting Lorentz force acting perpendicularly to both the magnetic field and current directions (i.e. in the y-direction) and the so-called Hall field arises:

$$E_y = \frac{I_x H_z}{ne} \tag{6.47}$$

where $I_x$ is the density of the current, $H_z$ is the intensity of the magnetic field and $n$ the density of carriers. The Hall coefficient is usually introduced,

$$R_H = \frac{E_y}{I_x H_z} = \frac{1}{ne} = \mu\varrho \tag{6.48}$$

where $\mu$ is the carrier mobility and $\varrho$ the conductivity of a given material. In this form, however, the relations shown above are valid only for materials

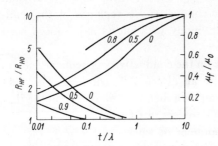

*Fig. 110:* Theoretical variation of the Hall coefficient and mobility with film thickness ($p$ is the parameter).

with one type of completely free carriers. The Hall constant varies with thickness for the same reasons the thin film conductivity varies. Theory affords complicated expressions which, however, reduce to simpler forms for extreme cases:

$$R_{Hf} = R_{HO} , \quad \gamma > 1 \tag{6.49a}$$

$$R_{Hf} = R_{HO} \frac{4}{3\gamma} \frac{1-p}{1+p} \frac{1}{(\ln 1/\gamma)^2}; \quad \gamma \ll 1, \quad p \ll 1 \tag{6.49b}$$

(The subscript $f$ denotes the quantities related to the film, 0 those related to the bulk.) Using (6.23) and (6.24) we obtain

$$\mu_f = \frac{\mu_0}{1 + \dfrac{3}{8\gamma}(1-p)}; \quad \gamma > 1 \tag{6.50}$$

$$\mu_f = \mu_0 \frac{1}{\ln(1/\gamma)} \; ; \quad \gamma \ll 1 \tag{6.51}$$

A plot of these relations is given in Fig. 110. The theory predicts an oscillatory behavior of $R_H$ at very large fields.

It is only recently that experimental results have been obtained which agree rather well with theoretical predictions. The curves measured on polycrystalline copper, for example, correspond to those in Fig. 110. Epitaxial films of copper, on the other hand, have exhibited a much weaker dependence, which may be attributed to the smooth surface of these films, resulting in a considerable specular contribution to the scattering. The oscillatory character of the $R_H$ dependence on magnetic field has also been experimentally verified on thin foils of Cd and Al at liquid helium temperature.

A further galvanomagnetic phenomenon we shall deal with is the additional resistivity of a conductor produced by a magnetic field. The paths of electrons are curved into a helix of radius $r$ given by

$$r = \frac{cmv}{He} \tag{6.52}$$

(where $c$ is the light velocity and all other symbols have the meaning mentioned above) and centered around the direction of the magnetic lines of force. If the electrons are really free, this effect cannot produce any changes in the bulk conductivity. The changes occur only when the electrons have a reduced degree of freedom due to binding. If, however, the electrons are scattered at the boundaries of the specimen, the field-induced deviations of their paths will result in a modified conductivity even in the case of completely free electrons. This fact enables much simpler relations for free electrons to be used in theoretical calculations even when some error is to be expected (as the term describing magnetoresistivity of the bulk is omitted from the result: electrons are indeed never totally free).

In order that an observable effect may appear it appears to be necessary that the radius $r$ be smaller than the mean free path and the latter, in turn, smaller than the specimen thickness. This makes rather rigorous demands on experimental conditions. If the effect is to be produced at attainable values of the magnetic field, it is necessary to work with materials in which electrons have a very long mean free path. The specimens used are monocrystalline films of rare metals in which the path is of the order of μm at liquid helium temperature.

182

The analysis of the experimental results frequently employs Kohler's rule:

$$\varrho_H - \varrho_0 = \varrho_0 f\left(\frac{H}{\varrho_0}\right) \tag{6.53}$$

where $\varrho_H$ and $\varrho_0$ are the conductivities with and without the applied field $H$ respectively and $f$ is a universal function.

The magnetoresistance may be one of two types: longitudinal or transverse, depending on the mutual orientation of the current and magnetic field.

Graphical illustrations of the theoretical results obtained by Kao for the longitudinal effect are shown in Fig. 111. Variation in the resistivity is plotted against the quantity $\beta = t/r$, $\gamma = t/\lambda$ being the parameter. It can be seen that with increasing $\beta$, the resistivity increases at first and then decreases until it approaches bulk resistivity at large values of $\beta$. The values for $\beta = 0$ agree with Fuchs's theory (see Sect. 6.2.1). For the position of the maximum, Kao's theory yields the relation

$$\beta_{\max} = 1.26\, \gamma^{0.57} \tag{6.54}$$

which may be used for calculation of the mean free path assuming the product $mv$ determining the radius $r$ is known (or vice versa). When

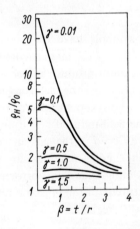

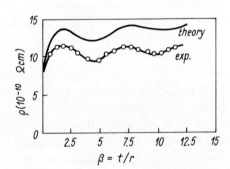

Fig. 111: Longitudinal magneto-resistivity of thin film as a function of $t/r$ (the parameter: $t/\lambda$).

Fig. 112: Resistivity variation with magnetic field for Al (0.081 nm) at 4.2 K in the Sondheimer arrangement (Amundsen and Olsen).

discussing experimental results, we have to keep in mind that the peak on the curve might have another cause. The magnetoresistance proper of the material (which is not considered in the theory of size effect) actually increases with field. This increase might at a certain point be just counterbalanced by the decrease in the size effect so that the maximum appears at that point. It is difficult in practice to find out which type a given maximum is.

The theoretical results presented thus far have been based on the assumption of completely diffuse scattering (subscript dif) at the boundary. If a fraction of electrons ($p$) are reflected specularly (subscript s) and if the fraction does not vary with the field, the relative change in film resistivity is given as

$$\left(\frac{\Delta\varrho}{\varrho}\right)_f = p\left(\frac{\Delta\varrho}{\varrho}\right)_s + (1-p)\left(\frac{\Delta\varrho}{\varrho}\right)_{dif} \qquad (6.55)$$

The value of $p$ may then be derived from a comparison of the experimental curve with the theoretical prediction.

Transverse magnetoresistance may be observed in two configurations: with the magnetic field either perpendicular ($H\perp$) or parallel ($H\parallel$) to the plane of the film.

In the case of $H\perp$, the theory elaborated by Sondheimer leads to the conclusion that the magnetoresistance variation with magnetic field shows some kind of oscillation (Fig. 112). This behavior is due to oscillations in the velocity of the electrons, which move in crossed electric and magnetic

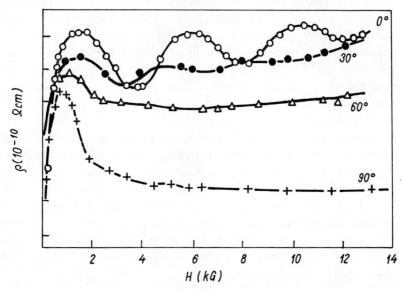

*Fig. 113:* The resistivity of Al at 4.2 K as a function of $H$ at various orientations of the field with respect to the normal plane of the film (Førsvoll, Hollweck).

fields. The effect does not occur in bulk material because the oscillating contributions from various depths of the specimen give a constant value in total. By limiting the film thickness the oscillations are brought to light as can be seen from the figure.

The second case, $H\|$, has been investigated by McDonald and Sarginson. According to their results, the resistivity variation with magnetic field should have only one maximum. Experiments confirm the prediction qualitatively but the field magnitude at which the maximum is observed is much higher than that of theory. The measurements obtained on films of Al are shown in Fig. 113. The lowest curve corresponds to $H\|$, the highest to $H\perp$, and the two middle ones correspond to two different orientations of the magnetic field with respect to the direction of the current.

During the study of the influence of orientation variations of the magnetic field on magnetoresistivity size effects in epitaxial films of certain metals (Ag, Au, Sn and In), remarkable anisotropic effects have been found. If the field direction coincides with some significant crystallographic orientations, very pronounced maxima or minima are observed. No theory of this anisotropy has yet been developed.

### 6.2.5 Superconductivity in Thin Films

The term superconductivity denotes the electrical state characterized by negligibly small values of resistivity. Such a state is established in some materials when their temperature drops below the critical temperature for the particular material. The current induced in a superconducting medium may circulate there for a practically indefinite period. It has been estimated from the time decay of such a current that the resistivity of superconducting material does not exceed $10^{-23}$ $\Omega$ cm, i.e. is $10^{17}$ times smaller than the resistivity of copper at room temperature. The critical temperatures are usually within several degrees Kelvin above absolute zero. The group of superconductors includes some metals (mostly those which have comparatively high resistivity at normal temperatures) as well as a number of compounds.

A significant feature of materials in a superconducting state is their ideal diamagnetism; they expel the force lines of a magnetic field so that the field penetrates to only a very limited depth. At a certain critical magnitude of an external magnetic field, which itself is a function of temperature, the superconducting state disappears.

The quantum theory of superconductivity has been developed in the past decade. From our point of view it is remarkable that the physics of thin films has played a substantial role in the elaboration of theoretical models and in the verification of their conclusions.

We shall at first present here some basic information on superconductivity in general and then we shall proceed to phenomena exhibited by thin films in the superconducting state.

The transition from the normal to the superconducting state is rather abrupt in the ideal case $(\Delta T \sim 10^{-3}\ \text{K})$ of very pure metals. In contaminated metals and in thin films the transition interval may be greater. The term critical in this case refers to the temperature at which the resistance falls to half the normal value.

The critical value of a magnetic field which destroys the superconducting state depends on temperature according to the relation

$$H_c = H(0)\left[1 - \left(\frac{T}{T_c}\right)^2\right] \tag{6.56}$$

where $H(0)$ is the critical field at $T = T_c$. The superconducting state may also be extinguished by passing a current through a superconductor which induces a magnetic field whose magnitude at the surface exceeds $H_c$.

Classical electrodynamics does not yield any satisfactory description of the phenomenon of superconductivity. The substitution of zero for resistivity in Maxwell's equations, for example, leads to the result that the time rate of change of magnetic flux approaches zero in a certain depth. This would imply that the magnetic field cannot disappear from the interior at the transition, which is at variance with observation.

The fact that there is apparently no scattering of electrons by the atoms of a crystalline lattice led to the conclusion that the wave functions describing the electrons in the superconducting state must be substantially different from those of the electrons in the normal state. In view of the fact that the lattice periodicity has no influence on the electrons, a hypothesis has been suggested that the wave function is not localized but has infinite spacial extent. In view of the uncertainty principle this implies a precisely defined momentum. The momentum of an electron moving in a magnetic field H, with the corresponding vector potential $A$, with velocity $v$ is given as

$$p = mv + \frac{eA}{c} \tag{6.57}$$

where $c$ is the velocity of light. If we introduce a current density $J$ into the equation we may rewrite it as

$$p = \frac{m}{ne}J + \frac{e}{c}A = \frac{e}{c}\left(\frac{4\pi\lambda_L^2}{c}J + A\right) \tag{6.58}$$

$$\lambda_L^2 = \frac{mc^2}{4\pi ne^2}$$

is the London penetration depth of the magnetic field. At normal levels of electron densities in superconductors the penetration depth is $\sim 10^{-6}$ cm. The observed values are somewhat higher $(\sim 5 \cdot 10^{-6})$.

If $p = 0$ then the corresponding wave function will be spatially limitless. From equation (6.58) the condition follows that

$$\frac{4\pi\lambda_{\mathrm{L}}^2}{c} J + A = 0 \qquad (6.59)$$

which inserted into Maxwell's equations results in

$$\Delta^2 H = \frac{4\pi n e^2}{mc^2} H = \frac{1}{\lambda_{\mathrm{L}}^2} H \qquad (6.60)$$

If we consider the boundary between a superconductor and surroundings as an infinite plane, the penetration of the magnetic field into the super-conductor will be described by

$$H(x) = H(0) \exp\left(-\frac{x}{\lambda_{\mathrm{L}}}\right) \qquad (6.61)$$

This circumstance may be significant, especially when the thickness of the superconductor is small, i.e., in the case of a thin film.

The temperature dependence of the London depth is given by

$$\lambda_{\mathrm{L}}(T) = \lambda_{\mathrm{L}}(0)\left[1 - \left(\frac{T}{T_c}\right)^4\right]^{1/2} \qquad (6.62)$$

London's deliberations led to the conclusion that a magnetic flux passing through a superconducting loop is quantized in units of $hc/q$ where $q$ is the carriers' charge. The experimentally found quantization demands we set $q = 2e$ to be in agreement with the theory. This fact seemed to indicate that the phenomenon of superconductivity involves a pair-wise electron transfer.

Although the theory of London was largely confirmed, certain discrepancies appeared, e.g. the dependence of experimental $\lambda_{\mathrm{L}}$ on impurities in superconductors and on the mean free path of electrons. This led to the elaboration of modified theories. Such theories have been proposed by Pippard, Ginzburg and Landau. They assume that the wave functions of electrons are extended only over a finite region and not over all of space and so achieve a satisfactory agreement with the experimentally observed dependences. This theory yielded a result of particular significance for thin film physics, namely, that the critical magnetic field required for annihilation of superconductivity increases with decreasing thickness of a sample.

This theory has also predicted the existence of hard and soft superconductors. The former preserve superconductivity in fields exceeding the critical one. The field penetrates in the separated filaments which represent quanta of magnetic flux. This 'mixed state' is characterized by simultaneous coexistence of superconducting and normal phases, the superconducting one forms layers of a thickness less than $\lambda_L$ and the normal one layers thicker than $\lambda_L$. Under certain conditions this state is preferred owing to the interface binding energy of both phases. The theory further predicted that the surface superconducting sheet would persist up to fields substantially exceeding the critical one.

There is a quantity which determines whether the transition will be abrupt (superconductors of type I) or gradual, occurring via an intermediate state (superconductors of type II). It is the Ginzburg-Landau parameter $\varkappa$, which is defined as the ratio of the London penetration depth $\lambda_L$ to the coherence range $\xi$, which depends on the purity of the material. Whether the given system will have a positive or negative surface energy depends on the magnitude of $\varkappa$. The transition occurs for $\lambda \approx \xi$, or, more precisely, for $\varkappa = \frac{1}{\sqrt{2}} = 0.71$. For $\varkappa < 0.71$ the surface energy is positive and the transition is of type I; for $\varkappa < 0.71$ the energy is negative and the type II transition occurs. Of the pure metals, only Nb, V and Tc have $\varkappa < 0.71$; in alloys one may even have $\varkappa \approx 100$.

The superconductors of type II behave as type I superconductors up to the field

$$H_{c1} = \frac{H_c}{\left(\varkappa \sqrt{2}\right)^{0.65}} \tag{6.63}$$

Above this field there arises the intermediate (or mixed) state, and, finally, on reaching the field

$$H_{c2} = \sqrt{(2)}\varkappa H_c \tag{6.64}$$

the material becomes a normal conductor. Thus type II superconductors may remain in the mixed state up to very high fields (of the order of $10^5$ G). However, even if $\varkappa \to \infty$ could be a reality, there is a threshold $H_p$ above which the mixed state cannot maintain itself: $H_p = 1.4 \cdot 10^6 \, T_c \, (A/m)$.

From the microscopic aspect the problem of superconductivity is solved by the theory of Bardeen, Cooper and Schrieffer (BCS). This theory maintains that a coupling is established between a pair of electrons with opposite spin and momentum through interaction with lattice phonons. The coupling is very weak indeed and consequently the energy difference between the normal and superconducting states is small. The existence of Cooper pairs is verified by experiments. In the energy spectrum the lower energies are situated at a distance $2\Delta$ from the normal states. This quantity represents a kind of forbidden band.

For the width of this band, theory provides the following expression:

$$\Delta = 2h\nu_L \exp\left(-\frac{1}{VN(\varepsilon_F)}\right) \tag{6.65}$$

where $\nu_L$ is the mean phonon frequency, $V$ the coefficient of electron-phonon interaction and $N(\varepsilon_F)$ is the density of states at the Fermi level. At $T = 0\,°K$ we have $2\Delta(0) = 3.52k \cdot T_c$, which agrees well with experiments. For $T = T_c$ one obtains $\Delta(T_c) = 0$. On account of the mentioned definiteness in the electron moments a strong correlation exists between individual pairs even at a very long distance, which leads to an array of remarkable effects. All pairs have the same energy (they obey the Bose-Einstein statistics) $E = h\nu$ and the total momentum $p = 2m\nu + 2eA$ (where $\nu$ is the velocity of a pair). The cloud of pairs, with its center of mass at $r$, may be thus described by a single wave function expressed as

$$\Psi = |\Psi| \cdot \exp(j\varphi) \tag{6.66}$$

where the phase given as

$$\varphi = \frac{Pr}{h} + 2\pi\nu t \tag{6.67}$$

is the same for all pairs. The absolute value of the phase is the same throughout the whole superconductor. The coherence of the phase gives rise to quantum effects which may be detected on a macroscopic scale. Every change of the phase in one part of a superconductor gives rise to a pair flux, which results in the re-establishment of the phase equilibrium.

If a superconductor surrounds a nonsuperconducting area threaded by a magnetic flux $\theta$, the flux must be quantized in order that the wave function be single-valued. This may be expressed by the Bohr-Sommerfield condition:

$$\int p \, dr = \oint (2m\nu + 2eA) \, dr = nh \tag{6.68}$$

A region may be found where $\nu = 0$. Then the condition holds for fluxes which are multiples of the magnetic flux quantum $\theta_0 = hc/2e$. This quantum is equal to $2.07 \cdot 10^{-15}$ Wb.

As regards superconducting thin films, it may be generally noted that their properties are markedly influenced by impurities and hence the method of their preparation is of paramount importance.

The BCS theory gives the following expression for the critical temperature:

$$k\,T_c = 1.14\hbar\omega_D \exp\left(-\frac{1}{V\,.\,N(\varepsilon_F)}\right) \tag{6.69}$$

where $\hbar\omega_D = k\theta_D$ ($\theta_D$ is the Debye temperature) and other symbols have the same meaning as in equation (6.65). Thus a change in $T_c$ may be effected through changes in the quantities occuring in (6.69). At first sight the influence of size is not evident from the relation; however, it does exist in connection with the rise of surface superconductivity.

Experiments have shown that the critical temperatures of thin films may differ by units of degrees from those of the bulk.

The soft superconductors were investigated first (e.g. Sn, Al, Pb), the hard ones later since their properties are more complex (e.g. W, Mo, To, Nb, V). Superconductivity was found down to extremely small thicknesses (of the order of 1 nm), when in fact individual islands instead of continuous films were present.

With decreasing film thickness, the critical temperature increases (Fig. 114) for some materials (e.g. In and Al), whilst for others a decrease has been observed (Pb), or a dependence with a maximum (Sn, Tl). The effects are sometimes explained as being due to the stress produced by cooling of the sample.

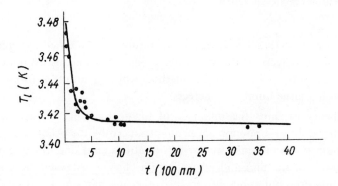

*Fig. 114:* Variation of critical temperature of indium with film thickness.

In a number of cases it has been found that a considerable increase occurs in the critical temperature of evaporated film, $T_{cf}$, compared to that of bulk, $T_{co}$ (e.g. $T_{co}$ for Al is 1.2 K, whereas $T_{cf} = 4.2$ K). In Bi there is practically no bulk superconductivity, whereas evaporated film is superconductive from 6 K. The superconductivity state persists up to 15 K, at

190

which temperature recrystallization and transition into the normal phase occur. A similar effect has been observed with beryllium.

The films considered here are produced by evaporation onto a cold substrate, thus consisting of very fine crystals or even exhibiting a completely amorphous structure; frequently the structures involved are not observed at higher temperatures. The film thickness has no influence here, nor has

Superconductive Properties of Thin Films of Some Metals .                    *Table 14*

| Material | $T_{cf}/T_{co}$ | $t$ (nm) | $T_{Debye}$ (K) | Deposition technique |
|---|---|---|---|---|
| Al | 2.6 | 100 | 370−400 | V |
| Mo | 5 | 10−400 | 375 | CS, EB |
| W | 400 | 10−400 | 315 | CS, EB |
| Sn | 1.26 | 50 | 110−210 | V |
| In | 1.22 | | 100−150 | V |
| Be | 800 | | 925 | V |
| Bi | 600 | | 111 | V |

V = evaporation onto cold substrate (vapor-quenched)
CS = cathode sputtering
EB = electron-beam evaporation

the substrate material. The critical temperature may be raised by intentional generation of defects in the structure, or by simultaneous evaporation of two components, or by any other method which guarantees formation of a film with a large number of defects.

Theory explains the influence of low-temperature condensation by the deformation of a phonon spectrum produced by the occurrence of nonequilibrium states in films. Similar deformations arise due to elongation or compression of the film and also during a change in crystallite sizes. The explanation does not apply universally; other mechanisms are also possible, e.g. the influence of changes in the electron spectrum in thin films and small crystallites, modifications of states near the surface, etc. The temperature $T_c$ is affected also by adsorbed layers and layers of compounds, e.g. oxides, formed on the surface.

Table 14 gives some data on superconducting thin films (according to Chopra). They reveal that the increase in the critical temperature is substantial in some cases and could perhaps have a practical significance.

Superconductivity may also be facilitated by establishing a multilayer system of different materials. A number of mechanisms has been proposed to account for these effects, yet none of them is wholly satisfactory.

It has already been noted that the critical magnetic field $H_{cf}$ increases with decreasing sample thickness. For very thin films $(t \ll \lambda_L)$ this dependence may be approximated as

$$H_{cf} = H_c \, 2 \, \sqrt{(6)} \frac{\lambda_L}{t} \tag{6.70}$$

and for thick films $(t \gg \lambda_L)$ as

$$H_{cf} = H_c \left(1 + \frac{\lambda_L}{t}\right) \tag{6.71}$$

Both relations hold for transitions of type I, i.e. when the normal-to-superconducting transition is sharp. This is the case for thick films $(t \gg \lambda_L)$ which have undergone thermal treatment. Thin films $(t \ll \lambda_L)$ with a highly disordered structure and large concentration of impurities belong to the group which exhibits a transition of type II, i.e. gradual decrease of resistivity with increasing magnetic field. Equation $(6.70)$ is also valid for this case. If, however, type II transition occurs in a film with $t \gg \lambda_L$ then the relevant equation is

$$H_{cf} = \sqrt{(2)}\varkappa H_c \tag{6.72}$$

where

$$\varkappa = \frac{\sqrt{(2)}qH_c\lambda_L^2}{\hbar c} \tag{6.73}$$

All we have said so far is valid in cases where the magnetic field is parallel to the film. When the field is perpendicular to the plane of the film then the critical field is given by $(6.72)$, i.e. all thin films behave as superconductors of type II.

The tunneling of electrons through a thin dielectric film which separates two metals, one or both of which are in a superconducting state, is an interesting process and contributed substantially to the verification of the theory of superconductivity. We shall deal later with the tunneling mechanism in the conductivity of thin dielectric films. Let us now only recall that tunneling is a quantum-mechanical phenomenon, that is, electrons are able to cross a region forbidden according to classical physical concepts, thanks to their wavelike nature, and to emerge on the other side of the barrier, provided there is an allowed and unoccupied energy level at their disposal (see also Sects. 6.2.2 and 6.2.6). For normal metals separated by a thin dielectric film (a few nm) theory predicts linear current-voltage characteristics (Fig. 115a) at low voltages. If one of the metals is in a superconducting state, a narrow forbidden gap appears at the Fermi level in the metal so that there are no levels allowed in the vicinity of the Fermi level. Consequently, no tunnel current is observed in current-voltage characteristics until the voltage is high enough to shift the allowed levels of the normal metal to correspond with those of the superconductor (Fig. 115b). The voltage at which the current starts to rise rapidly corresponds to a halfwidth of the forbidden band of the superconductor, and the halfwidth may be determined in this way. If there is the same superconductor on both sides, it is necessary to apply the voltage corresponding to the whole width of the forbidden band

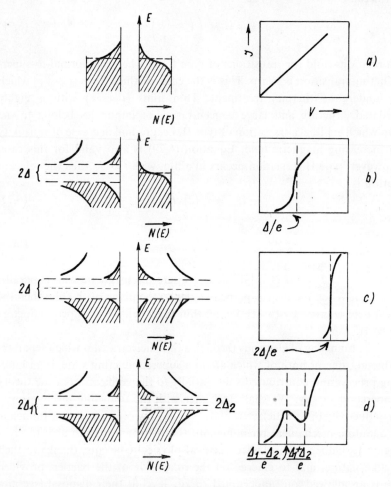

*Fig. 115:* Characteristics of tunneling: (a) between normal conductors; (b) between superconductor and normal one; (c) between two identical superconductors; (d) between two different superconductors; on the left are the density-of-states diagrams, on the right the current-voltage characteristics.

before the tunnel current starts flowing (Fig. 115c). If the electrodes are made from two different superconductors with forbidden bands of width $2\Delta_1$ and $2\Delta_2$, respectively, a negative-resistivity region may arise (Fig. 115d). Up to the voltage $(\Delta_1 - \Delta_2)/e$ the current increases because there are free states available for the tunneling electrons. With increasing potential, the current begins to decrease owing to the decrease in the density of the available states. Only after the voltage reaches $(\Delta_1 + \Delta_2)/e$ will the occupied states of one of the metals be set opposite the empty allowed states of the other and the current will again start to rise rapidly.

The effects just mentioned have afforded a direct experimental proof of the existence of a forbidden band in superconductors and enabled its width to be determined. In addition, the system illustrated in Fig. 115d may, as any other system with a negative-resistivity region, serve for the generation of electrical oscillations.

Besides normal tunneling, there is, according to the theory of the superconducting state, there is the possiblity of tunneling of Cooper pairs provided the film is very thin ($\leqq$ nm) and the magnetic field very weak ($\leqq$ several gauss). This so-called Josephson tunneling does not generate any potential fall up to a certain current intensity (Fig. 116).

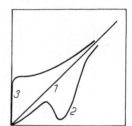

Fig. 116: Current-voltage characteristics of: (1) normal, (2) superconductive, (3) Josephson tunneling.

If the barrier separating two superconductors is rather thick the phases in the material are independent, say, $S_1$, $S_2$. If the distance is small ($d_c \leqq$ $\leqq 2$ nm), the interaction energy between the pairs may be greater than $kT$ (i.e. the pairs are no longer easily destroyed by phonons) and a current ensues which counterbalances the phase difference $\varphi' = \varphi_1 - \varphi_2$. This is called weak coupling:

$$I = I_c \sin \varphi' \tag{6.74}$$

The critical current is determined by parameters of the barrier, its dimensions and electrical properties. The dependence of this current on the magnetic field is typical:

$$I = I_c \frac{\sin (\pi \Phi_s / \Phi_0}{\pi \Phi_s / \Phi_0} \tag{6.75}$$

where $\Phi_s$ is the magnetic flux through the nonsuperconducting region, $\Phi_0$ is the magnetic flux quantum which has been defined above. The dependence of the current on the magnetic field parallel to the plane of a film is shown in Fig. 117a. This variation has indeed been observed. It is evident

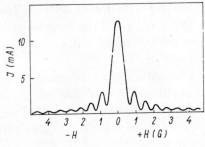

*Fig. 117a:* The dependence of the Josephson current on the magnetic field (Sn/Sn) (Fisher, Giaever).

from the figure that very small magnetic fields reduce the current substantially. The presence of the terrestrial magnetic field, which reduces the current to about $1/100$, is one of the reasons why the effect had not been discovered sooner.

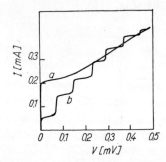

*Fig. 117b:* Current-voltage characteristics of a Josephson junction (weak coupling): (a) without external field, (b) with external magnetic field.

The form of the dependence in Fig. 117a is suggestive of Fraunhofer's diffraction effect in optics. The interference effects may be made more pronounced by connecting several such thin junctions into a parallel sequence.

All this refers to the Josephson effect which appears if there is no potential fall across a thin junction, i.e. as long as the current remains subcritical. If the current exceeds the critical value, a potential $V$ is developed across the junction which, according to Josephson, gives rise to a time variation of the wave-function phase difference between both superconductors. The current of Cooper pairs will oscillate with angular frequency

$$\frac{d\varphi'}{dt} = \omega_j = \frac{2eV}{\hbar} \qquad (6.76)$$

The ratio of the frequency to the voltage difference across the barrier is very large: $-483.6\,\text{MHz}/\mu\text{V}$. The transfer across the barrier involves an energy change of $\Delta E = 2eV$, i.e. the angular frequency $\omega_j = \Delta E/h$. Thus, quantum transitions occur between the states $E_1$, $E_2$ accompanied by emission of electromagnetic radiation. The radiation is monochromatic and, owing to correlation of phases, coherent (the quanta are emitted with identical phases). The system thus represents a new kind of quantum generator. The recorded radiated power has so far been very low $(10^{-8} - 10^{-12}\,\text{W})$ but this is due to an inability to adapt the generator with the detector to a sufficient degree. It may be expected that energies of the order of $10^{-2}\,\text{W}$ will be obtained in this way in the future.

On interaction with an external electromagnetic field, the current-voltage characteristic exhibits jumps whenever the frequency of the external generator equals a multiple of the intrinsic frequency: $\omega_g = N \cdot \omega_j$ (see Fig. 117$b$).

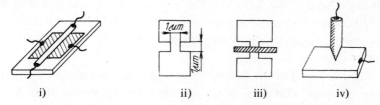

*Fig. 117c:* Various types of Josephson junctions: (i) sandwich; (ii) bridge; (iii) bridge with nonsuperconductive overlayer (iv) point junction.

A classical example of a Josephson junction is the sandwich structure (Fig. 117$c$) made up of two superconductors separated by a thin dielectric film. Direct superconducting short-circuits between the metals are very detrimental in this configuration. On the other hand, thickness irregularity and surface roughness only lead to deterioration of the junction parameters. Besides the type denoted usually as S—I—S (superconductor—insulator—superconductor) other configurations are used, namely, S—N—S (N—nonsuperconducting metal) or S—c—S, where c denotes a narrowed region of the superconductor, i.e. a kind of bridge (see Fig. 117c—ii). The last type is usually further modified by inserting a layer of N-material across the bridge; this weakens the interaction between the superconductors and thus decreases an undesirable temperature dependence (Fig. 117c—iii). For the sake of completeness we should note that junctions are also made in the form of tips placed often with variable pressure, on a substrate (Fig. 117c—iv). The sandwich structure, developed, for example, in the form Nb—Nb—O—Nb, has the disadvantage of having a relatively large capacity and is thus not suitable for use in high-frequency circuits.

Superconductivity in general and both Josephson effects in particular have recently found numerous applications to which we shall return in Sect. 7.2.5.

### 6.2.6 Conductivity of Thin Dielectric Films

The term dielectric denotes a material which at normal temperatures has a very small number of free charge carriers. The dielectrics have a broad forbidden band (of the order of several eV) so thermal excitation alone is not able at normal temperatures to excite electrons from the valence to the conduction band. Nevertheless, currents of very large densities may pass through a dielectric, especially through thin dielectric films, when a sufficient number of free carriers are injected from the electrodes into the conduction band of the dielectric. The conductivity of thin dielectric films (TDF) is examined mainly in the configuration in which a TDF is inserted between two electrodes. Such an arrangement is often called a sandwich structure.

Before we start to describe the particular mechanism of current passage through TDF, we should mention one important fact. A considerable part of the experimental work on TDFs has been done on films with an amorphous character. Typical examples of such films are the oxides prepared by thermionic or anodic oxidation on surfaces of some metals such as aluminium, tantalum, zirconium, etc. Now, the band theory of solids is based on the assumption of a periodic crystal lattice. The question arises, therefore, whether we may apply the results of this theory, namely band structure, without modifications to amorphous materials. The problem has recently been studied in a number of theoretical papers. It has been shown that the fundamental feature, i.e. the way in which the allowed energies are divided into defined bands, remains principally preserved because of the fact that it depends above all on the interaction between neighboring atoms, which is the same in an amorphous material as in a crystalline one. Owing to irregularities in the lattice, however, the limits of bands become somewhat foggy and a tail of allowed states appears in their vicinity (Fig. 118). If the tails from the valence and conduction bands overlap, the material assumes a semiconductive rather than a dielectric character. A remarkable fact has been found experimentally, namely, that in contrast to normal semiconductors the conductivity of these amorphous structures is only weakly sensitive to impurities. The explanation is seen in large fluctuations in local arrangements of atoms so that a large number of localized trap levels exists which may capture free carriers and so compensate the effect of impurities.

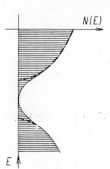

*Fig. 118:* Energy diagram of an amorphous substance.

Amorphous substances are mostly characterized by a very small mobility of carriers. If this quantity is less than about $5\,\text{cm}^2/\text{Vs}$, the effective mean free path of an electron is in turn shorter than the lattice constant, which is patently impossible and indicates that the concept of mean free path cannot be used here. We have to bear all this in mind when interpreting phenomena exhibited by dieletric films.

It has already been noted that conductivity may be artificially induced in dielectric films by injecting free carriers into them. In thin films of ionic crystals, a typical example being the alkali halides, an additional contribution comes from the ionic mechanism of charge transfer. The crystal lattice is never perfect and it always contains vacancies, i.e. points unoccupied by ions. An applied electric field (especially at elevated temperatures when the amplitudes of thermal vibrations of lattice points are large) facilitates jumps of ions into the adjacent vacant sites, and in this way an ionic current is established.

The lattice potential is more or less periodic so that a potential barrier exists between two neighboring crystal lattice sites. If $\Phi$ is the barrier height, then the jump frequency (without the electric field) is given as

$$\bar{v} = \frac{1}{bv^2}\left(\frac{kT}{\hbar}\right)^3 \exp\left(-\Phi/kT\right) \qquad (6.77)$$

where $v$ is the vibrational frequency in directions perpendicular to the jump, $b$ is the number of possible jump directions, and the other symbols have their usual meaning. If no electric field is present, the jumps occur in random directions and the resulting current is consequently zero. If, however, an electric field of intensity $F$ is applied along the $x$-direction, then the potential barrier is lowered in one direction and heightened in the opposite by $eFl$, where $l$ is the distance between the sites. The jump frequencies in these directions $(+F$ and $-F)$ will be

$$\bar{v}_{+F} = \frac{1}{bv^2}\left(\frac{kT}{\hbar}\right)^3 \exp\left(-\Phi/kT\right)\exp\left(+Fel/2kT\right) \qquad (6.78\text{a})$$

$$\bar{v}_{-F} = \frac{1}{bv^2}\left(\frac{kT}{\hbar}\right)^3 \exp\left(-\Phi/kT\right)\exp\left(-Fel/2kT\right) \qquad (6.78\text{b})$$

The total probability of a charge transfer is given by the difference

$$\bar{v}_\text{P} = v_{+F} - \bar{v}_{-F} = 2\,\bar{v}\sinh\frac{eFl}{2kT} \qquad (6.79)$$

If the field is so weak that $eFl \ll kT$ then we may write an approximate expression:

$$\bar{v}_\text{P} = \bar{v}\,\frac{Fel}{kT} \qquad (6.80)$$

The current density is given as

$$i = ne\bar{v}_\text{p}l = n\bar{v}\frac{Fe^2l^2}{kT} \qquad (6.81)$$

where $n$ is the concentration of charge carriers; thus the dependence on the field is linear (Ohm's law applies).

If, on the other hand, the field is very strong $(\geqq 10^5 \text{ V/cm})$, i.e. $eFl > > kT$, the probability of backward jumps is so small that it may be neglected. Then

$$i = n\,ev_{+F}\,l = i_0\exp\left(-\frac{\Phi}{kT} + \frac{Fel}{2kT}\right) \qquad (6.82)$$

Experiments reveal, however, that with very powerful fields even $l$ undergoes a change. The main difficulty in the experimental verification of these relations lies in separating the ionic currents from those of electrons. The characteristic feature for the ionic mechanism of charge transfer is simultaneous transport of mass. Further, polarization effects often appear due to formation of space charges in the film; these result in an uneven distribution of potential and eventual decreases of current. Moreover, the activation energies typical for ion transfer are considerably greater than those for the pure electron mechanism (several eV, in contrast to less than 1 eV for electrons).

In spite of all these differences, it is in some cases extremely difficult to separate both mechanisms since the space charges and the polarization effects sometimes arise also due to electron current, while activation energies and mobilities may also be comparable.

The energy diagram in Fig. 119 illustrates various possibilities of charge transfer in thin dielectric films. If the film is very thin (several nm) and the applied voltage is not too high, the electrons may flow by tunneling

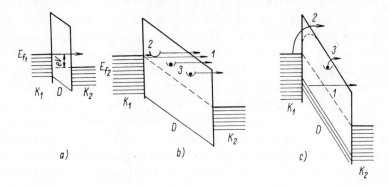

*Fig. 119:* The energy diagrams of possible mechanisms of charge transport in a dielectric film D: (a) tunneling between electrodes $K_1$ and $K_2$; (b) tunneling: 1 — into conduction band; 2, 3 — from traps; (c) 1 — Zener breakdown; 2 — Schottky effect; 3 — Poole-Frenkel effect; $E_{f1}$, $E_{f2}$ — Fermi levels in $K_1$, $K_2$ metals.

from the negative electrode into unoccupied levels of the positive electrode (Fig. 119a). If the voltage is higher or the film is thicker so that $Ft > \Phi$, the tunnel transfer conveys the electrons into the conduction band of the dieletric rather than directly to the second metal (Fig. 119b, transition 1). If the dielectric contains a large number of traps, the tunneling can take place through some of the traps (2). The calculation shows that the transfer probability may be increased substantially in this way. The tunneling can also occur from some occupied trap into the conduction band (3), or (at high applied field) from the valence band into the conduction one (Fig. 119c, transition 1). This mechanism is sometimes called the Zener's breakdown. All mechanisms mentioned here are basically tunnel transfers and, as we shall see later, are governed by analogous laws.

Besides these mechanisms, electrons may be injected into the conduction band of the dielectric by processes which are induced by thermal activation energy, namely thermionic emission of electrons from the negative electrode into the conduction band of the dielectric facilitated by the lowering of the barrier by an applied electric field (Schottky effect, Fig. 119c, transition 2) and the thermal ionization of traps, which is also facilitated by a strong electric field (the Poole-Frenkel effect, transition 3).

All the processes discussed feed free carriers into the conduction band of the dielectric. The carriers undergo various interactions in the dielectric and thus have a limited mean free path. They lose energy by relatively small quanta in interacting with the optical phonons (i.e. with the vibrations of the crystal lattice) or they may be captured by traps. They are accelerated during their travel by an existing electric field (this is shown in the diagram Fig. 119*b* by increasing the distance from the bottom of the conduction band). The electrons trapped in the dielectric constitute a space charge which may substantially affect the conditions for further charge transfer. This is especially evident when the number of injected electrons is large and does not itself represent the limiting factor. This phenomenon occurs when the contact of the metal with the dielectric is such that the boundary barrier is slight and the electrons can travel into the dielectric very easily. This is the case of ohmic contact. We then speak about space-charge-limited currents. It may occur in materials with a high density of defects where localized levels are so near to each other that the tunnel or thermal transition of electrons may take place from one level to an adjacent one. We then speak about hopping mechanism of conduction.

Let us now consider some of the mechanisms in more detail.

(*a*) The tunneling of electrons through a thin barrier between two metals. The first case for which the relevant quantum mechanical relations were derived was that of a rectangular barrier (Fig. 120a). There is a certain probability of $1 \to 2$ electron transfer and also of reverse transfer. The

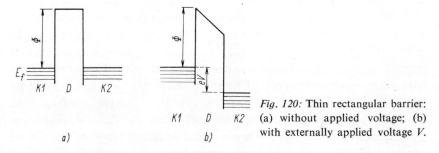

*Fig. 120:* Thin rectangular barrier: (a) without applied voltage; (b) with externally applied voltage $V$.

resultant current (which will be different from zero provided some voltage is applied across the barrier, Fig. 120b) will be equal to the algebraic sum of both currents:

$$i = \frac{4\pi m e}{\hbar^3} \int_0^\infty (f_1 - f_2)\, \mathrm{d}E \int_0^{E_{max}} D(E_x)\, \mathrm{d}E_x \qquad (6.83)$$

Here $f_1$ and $f_2$ are the energy distribution functions of electrons in the first and second metals respectively, $D$ is the barrier transparency (i.e. the tunneling probability for an electron of energy $E$ where $E_x$ corresponds to the component of momentum perpendicular to the plane of the film). The determination of the transparency is the pivotal problem of the whole theory because this function in turn determines the current-voltage characteristic of the system. Quantum mechanics enables the problem to be solved in a quasiclassical approximation (the Wentzel-Kramer-Brillouin method-WKB):

$$D(E_x) = \exp\left[ -\frac{4\pi}{\hbar} \int_{s_2}^{s_1} \sqrt{(-p_x^2)}\, dx \right] \tag{6.84}$$

where $p_x$ is the $x$-component of momentum, and $s_1$, $s_2$ are the points which delimit the effective thickness of the barrier, i.e. intersections of the potential curve of the barrier with the Fermi level. Assuming the classical relation between $p_x$ and $E_x$, i.e.

$$E_x = \frac{p_x^2}{2m} \tag{6.85}$$

we may write (6.74) in the form

$$D(E_x) = \exp\left[ -\frac{4\pi}{\hbar} \int_{s_1}^{s_2} \sqrt{\{2m[\Phi(x) - E_x]\}}\, dx \right] \tag{6.86}$$

$\Phi(x)$ is the barrier height above the Fermi level.

By substitution in (6.83) and subsequent integration we obtain an explicit expression only in the simplest cases. For a rectangular barrier and a very small applied voltage we obtain

$$i = \frac{e^2 (2m\Phi)^{1/2}}{\hbar^2 s} V \exp\left[ -\frac{4\pi s}{h} (2m\Phi)^{1/2} \right] \tag{6.87}$$

For very large voltages, i.e. for high fields, $f = V/(s_2 - s_1)$, we obtain

$$i = \frac{e^2 F^2}{8\pi h \Phi} \exp\left[ -\frac{8\pi}{3heF} (2m)^{1/2} \Phi^{3/2} \right] \tag{6.88}$$

The theory has been extended chiefly by Stratton and Simmons so that it is possible to determine the current passing through a barrier of arbitrary profile and at arbitrary temperature (equations (6.87) and (6.88) hold when no electrons are excited above the Fermi level, i.e. strictly taken only at the temperature of absolute zero). It has been demonstrated that an arbitrary barrier may be replaced by a rectangular barrier of thickness $\Delta s = s_2 - s_1$ and of height given by the mean height

$$\bar{\Phi} = \frac{1}{\Delta s} \int_{s_1}^{s_2} \Phi(x)\, dx \tag{6.89}$$

The density of transmitted current is then determined for all values of the applied voltage $V$ by the relation

$$i = \frac{6.2 \times 10^{10}}{(\Delta s)^2} \{ \bar{\Phi} \exp \left( -1.025 \, \Delta s \bar{\Phi}^{1/2} \right) -$$

$$- (\bar{\Phi} + V) \exp \left[ -1.025 \, \Delta s \left( \bar{\Phi} + V \right)^{1/2} \right] \} \qquad (6.90)$$

The concrete numerical values of the constants have been used here $\bar{\Phi}$ is in eV, $\Delta s$ in Å and the current density in A/cm$^2$. It should be noted that the actual shape of the barrier is usually complex, being affected by image forces and by space charge in the dielectric. The shape is not usually known exactly and various approximations are used instead.

The temperature dependence has been derived by developing the relevant functions into series. It is expressed by the relation

$$\frac{i(V, T)}{i(V, 0)} = \frac{\pi BkT}{\sin \pi BkT}; \quad B = f(\bar{\Phi}) \qquad (6.91)$$

where $i(V, T)$ is the current density at a temperature $T$ and $i(V, 0)$ the density at the temperature of absolute zero. This expression may also be developed into a series. The temperature dependence is usually described with sufficient accuracy by the first term of the series. By substituting numerical factors we obtain

$$\frac{i(V, T)}{i(V, 0)} = 1 + \frac{3 \cdot 10^{-9} (\Delta s T)^2}{\bar{\Phi}} + \ldots \qquad (6.92)$$

Experiments in this respect have mostly been carried out on systems of very thin films of oxides prepared by oxidation of a metal, particularly the system Al—Al$_2$O$_3$—Al (other metals may be also used for the electrodes). With regard to the functional dependence, very good agreement with theory has been attained. Comparison of absolute values is difficult since the basic parameters of the barrier are not known with sufficient precision. A numerical agreement between theory and experiment is frequently assumed and the parameters of the barrier (especially its thickness) are then determined. If the dielectric thickness obtained from capacitance measurements is substituted for the barrier thickness and the geometric area of the electrodes is taken as the barrier area, the resulting current densities are lower than the measured ones by several orders. The most probable cause for the discrepancy is to be found in irregular thickness of the dielectric. The current depends considerably on this quantity, and in practice the current only flows through the thinnest regions so that the actual area through which the current passes may even be smaller than the geometric one by two orders of magnitude.

In addition to our ignorance of the actual barrier shape, no reliable knowledge exists concerning the effective mass of electrons and the dielectric constant of the film. On very thin films and with fields a penetration of the field into the electrodes may even occur. If electrons do not tunnel directly

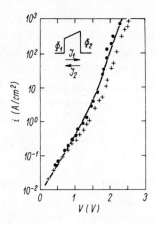

*Fig. 121:* Current-voltage characteristics of the system Al—Al$_2$O$_3$—Al: + is for $I_1$ at 300 K; ● is for $I_2$ at 300 K, the curve is $I_2$ when $\Delta s = 1.75$ nm, $\Phi = 1.5$ eV, $\Delta\Phi = 0.35$ eV, $\varepsilon = 12$ (Pollack-Morris).

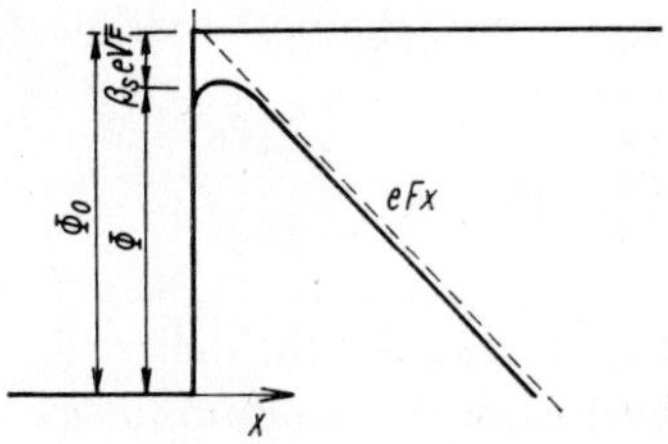

*Fig. 122:* Energy diagram of metal-vacuum interface illustrating the Schottky effect.

into the other metal but drift instead into the conduction band of the dielectric, the justification for the use of (6.90) becomes doubtful and the interaction of the electrons with the dielectric can no longer be neglected.

An example of the current-voltage characteristics of the Al—Al$_2$O$_3$—Al system is given in Fig. 121 (according to Polack and Morris). It is seen that the current densities attain considerable values. Experiments reveal that the temperature dependence is greater than the theoretical one. The results may, however, be affected markedly by temperature-effected changes in parameters which occur in the fundamental equation, e.g. the temperature dependence of the barrier height.

For liberation of electrons from traps or from a valence band of a dielectric relations similar to (6.88) hold at high intensities of electric field provided a proper barrier height is substituted. (The first factor usually has another structure but this does not significantly modify the characteristics.)

(b) Thermionic emission over the barrier lowered by a strong electric field (Schottky effect) is governed by the Richardson-Schottky relation.

The density of the thermionic current from a material with work-function (height of surface barrier) $\Phi$ is given as

$$i = AT^2 \exp\left(-\frac{\Phi}{kT}\right) \tag{6.93}$$

where $A$ is Richardson's constant. Its theoretical value is 120 A/cm$^2$ × × degrees, but, owing to the temperature dependence of the work-function, it actually assumes values ranging from a few to several hundreds of A/cm$^2$. . degrees. In the presence of a strong electric field the profile of the barrier of the initial height $\Phi_0$ is modified and its height is lowered (Fig. 122). The dependence of the potential on the distance $x$ from a surface is

$$\Phi = \Phi_0 - eFx - \frac{e^2}{16\pi\varepsilon\varepsilon_0 x} \tag{6.94}$$

where $\varepsilon$ and $\varepsilon_0$ are the dielectric constants; for the maximum, which can be derived from the condition $\partial\Phi(x)/\partial x = 0$, we obtain

$$\Phi_{\max} = \Phi_0 - \frac{1}{2}e\sqrt{\left(\frac{eF}{\pi\varepsilon\varepsilon_0}\right)} = \Phi_0 - \beta_s eF^{1/2} \tag{6.95}$$

where $\beta_s$ is the Schottky's constant.

Thus the density of the thermionic current will be given as

$$i = AT^2 \exp\left(-\frac{\Phi_0 - \beta_s eF^{1/2}}{kT}\right) = i_0 \exp\left(\frac{\beta_s eF^{1/2}}{kT}\right) \tag{6.96}$$

The current in this case increases exponentially with the square root of the applied voltage. In contrast to the tunneling mechanisms, there is now a very pronounced exponential dependence of the current on temperature.

Schottky's currents are significant especially for lower barrier heights and higher temperatures. The behaviour corresponding to equation (6.96) has mainly been observed on somewhat thicker films (more than 10 nm).

It is experimentally difficult to separate the Schottky current from that resulting from the Poole-Frenkel effect. The thermal ionization of traps, which is also facilitated by an electric field, is governed by a similar law:

$$i = i_0 \exp\left(\frac{\beta_{PF} eF^{1/2}}{kT}\right) \tag{6.97}$$

The difference is in the constant,

$$\beta_{PF} = 2\beta_s \qquad (6.98)$$

However, when the constant is inferred from the slopes of lines in the plot of $\ln i \sim f(\sqrt{V})$, the resulting values approach the Schottky value even when the process involved is, with a great probability, bulk-dependent and thus determined by the Poole-Frenkel effect.

Various explanations have been proposed for this discrepancy. The most plausible is that which takes into account that the potential representing the trap is deformed by the field so that the escape of electrons is made easier in one direction and more difficult in the opposite one. When the probability of electron transfer through such an asymmetrical barrier is then calculated, the resultant dependences may in effect approach Schottky's function. To distinguish these effects unambigously is difficult, and physical considerations rather than the observed quantitative differences are used for that purpose.

Since the experimental work has chiefly been done on amorphous materials, the possibility should be considered that electrons are transferred into the quasicontinuum of states constituting a tail at the bottom edge of the conduction band instead of being directly transferred into the conduction band. The activation energy pertinent to this transfer will be, of course, different.

An example of current-voltage characteristics corresponding to Poole-Frenkel's (or Schottky's) effect is given in Fig. 123.

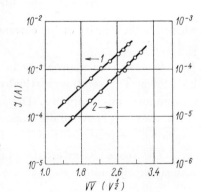

Fig. 123: Current-voltage characteristics of Pb—$Al_2O_3$—Pb system: (1) $T = 273$ K; (2) $T = 241$ K (Pollack).

(c) The laws governing the currents limited by space charge differ according to whether the current is conveyed by one or two kinds of carriers (i.e. only by electrons, holes, or by electrons together with holes) and whether the dielectric does or does not contain traps. As to the first criterion, the situation depends on the nature of the contacts. If only one of them is ohmic, then only one group of carriers is present; if both the anode and cathode form ohmic contacts, both kinds of carriers occur in the dielectric. We speak then about single or double injection. We assume that one or both contacts supply a sufficient number of free carriers so that the current is not limited by the number of available carriers but by the space charge present. The charge is constituted in part by that of the free carriers and in part by the trapped charge.

In all cases the reasoning proceeds from the solution of the fundamental equations, namely, the equation for the current density

$$i = e(\mu_n n + \mu_p p) F - e\left(D_n \frac{dn}{dx} - D_p \frac{dp}{dx}\right) \tag{6.99}$$

where $\mu_n$ and $\mu_p$ are the mobilities of electrons and holes respectively, $n$ and $p$ the densities of the injected carriers, $D_n$ and $D_p$ the diffusion coefficients for electrons and holes, respectively (in some cases the diffusion current is neglected). Further, Poisson's equation is introduced (for the one-dimensional case)

$$\frac{\partial F}{\partial x} = \frac{\varrho}{\varepsilon} \tag{6.100}$$

where $\varrho$ is the density of the space charge and $\varepsilon$ is the dielectric constant. Finally the equation of continuity is added,

$$\text{div } i = 0 \tag{6.101}$$

where $i$ is the density of the current conveyed in general by both electrons and holes.

In each case appropriate boundary conditions should be explicitly laid down. The solution is usually carried out under certain simplifying assumptions on computers. Derivation of the currrent-voltage characteristics would go beyond the scope of this book. Only some of the important results will be presented here (see [7], [17]).

The space-charge-limited current in a dielectric without traps and with single injection is given as

$$i = \frac{9}{8} \varepsilon\mu \frac{V^2}{t^3} \tag{6.102}$$

where $\varepsilon$ is the dielectric constant of the dieletric, $\mu$ is the drift mobility of charge carriers, $V$ the applied voltage, and $t$ the thickness of the dielectric.

If shallow traps are present (trap levels are situated between the Fermi level and the bottom of conduction band) then

$$i = \frac{9}{8} \vartheta \varepsilon \mu \frac{V^2}{t^3} \tag{6.103}$$

where $\vartheta$ is the ratio of the density of electrons in the conduction band to the density of trapped electrons. This coefficient may have a very small value (down to $10^{-7}$), which means that the presence of the traps may radically reduce the current. The deep traps which lie below the Fermi level do not affect the current since they are always filled up and thus cannot take part in the capture of the injected carriers.

If more kinds of traps are present, the laws are more complex. At double injection but without the presence of recombination centers, the current-voltage characteristics assume a similar form:

$$i = \frac{9}{8} \mu_{ef} \varepsilon \frac{V^2}{t^3} \tag{6.104}$$

where $\mu_{ef}$ is given by a rather complicated function of the drift mobilities of both types of carriers, and at small values of these mobilities may be $10^3$ times larger. Thus the occurrence of both kinds of carrier may increase the current substantially, which is the consequence of mutual neutralization of the corresponding space charges.

If, in addition, recombination takes place at recombination centers, the current-voltage characteristics may be written as

$$i = \frac{9}{8} \varepsilon \mu_n \mu_p \tau_h N_r \frac{V^2}{t^3} \tag{6.105}$$

where $N_r$ is the density of recombination centers and $\tau_h$ the mean life-time of carriers at high injection levels (i.e. at high densities of the injected currents). This relation holds if the recombination centers have a much greater effective cross-section for holes than for electrons and the effect of the space charges is neglected.

For double injection, the characteristics, may, under certain conditions, exhibit a region of negative resistance. In this region, the potential fall between the contacts decreases with increasing current and thus the characteristic assumes a sort of $S$ shape from which it derives its type-name (Fig. 124). The effect is due to a sharp increase in conductivity which may be caused by the dependence of the carriers' life-time on the level of injection.

We have cursorily reproduced some of the results concerning the space-charge-limited currents in dielectrics in order to indicate the complexity

208

of this topic. Much of the experimental work devoted to the problem was the result of the attempt to construct solid-state electronic elements that would replace vacuum tubes. As we have seen, the dielectric diode with one injecting contact works similarly to the vacuum diode: it exhibits a rectifying effect (with a rectification coefficient up to $10^6$), and it can transmit con-

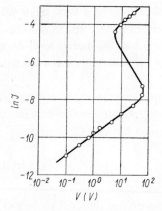

*Fig. 124:* Current-voltage characteristics of space-charge-limited current (negative resistance-S-shape).

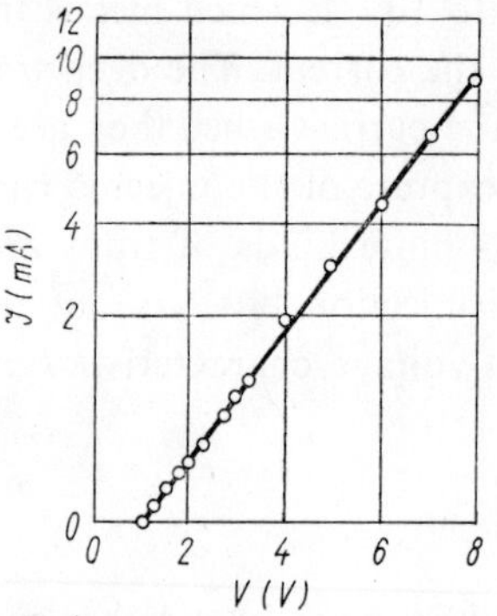

*Fig. 125:* Characteristic of space-charge-limited current (In—CdS—Au).

siderable currents and operate at higher temperatures than the semiconductor rectifiers. In this respect the experiments have mainly been conducted with such materials as CdS, CdSe, ZnS, SiC, $Al_2O_3$, etc. The samples prepared by evaporation *in vacuo* or by a similar technique have notable concentrations of traps $(10^{14} - 10^{20}$ cm$^{-3})$. Relatively thick films have usually been used (of the order of μm) and experiments have also been made on single crystals. A great diversity of results has been obtained probably owing to the diversity in actual structures of individual samples. In some cases the experimental results accord well with the theory (see, e.g., Fig. 125); in other cases there are considerable discrepancies: the parameters of films determined on the basis of the theory do not agree with the same parameters determined independently by other methods (e.g. trap concentration).

(*d*) Conductivity through centers of impurity is mainly observed in dielectrics or in semiconductors at very low temperatures, when it is not obscured by conductivity due to electrons in the conduction band. It is characterized by very low effective mobility of carriers since the electrons hop between adjacent centres in which they stay for a certain time. This is also true for the holes. Depending on the concentration of trap centers, conduction may occur in two possible ways: If the impurity concentration is relatively low $(10^{15}$ to $10^{17}$ cm$^{-3})$, the captured electrons are localized in individual centers (in practice their wave functions do not overlap). A certain probability exists of electron transfer from one center to an adjacent empty one either by tunneling or by transfer across the barrier due to thermionic excitation (hopping). The presence of an electric field will facilitate transfers in the field direction and thus an electric current will arise. The current depends strongly on the impurity concentration.

In addition, a certain degree of compensation is needed before the current starts to flow, i.e. donor centers must be compensated in part by the presence of acceptor levels into which some of the electrons may be transferred and so clear the way for the transfer between the donor centers. This type of conduction is characterized by a small activation energy, linear current-voltage characteristics and by conductivity increasing with frequency.

If the impurity concentration is high, the wave functions of electrons in individual centers overlap considerably so that the centers may give rise to a continuous impurity band. The resultant conductivity has then a metallic character.

This brief survey of some mechanisms of electron transport in thin dielectric films is by no means exhaustive. Additional effects exist which we have not mentioned. Nevertheless, we may conclude that on the whole the transport phenomena in dielectrics are not yet completely understood, that the theory rests on models, that are too simple and that experimental results are often of dubious significance owing to uncertainty as to the actual conditions in the systems examined. Since the phenomena are of great importance for electronic applications a large amount of additional effort will certainly be devoted to them.

### 6.2.7 Dielectric Properties of Thin Films

Since large values of capacity in a small volume may be produced by using thin dielectric films, it is advisable to study their dielectric properties, namely, the dielectric constant or permittivity, loss angle (tg $\delta$) and dielectric strength (breakdown voltage).

The first question is whether thin films possess the same dielectric constant as bulk materials or whether the quantity is thickness-dependent. From calculations it follows that the magnitude of the constant should be preserved until an extremely small thickness of several monolayers is reached. This result has so far been verified by experiment in only one case, that of cadmium stearate, where the dielectric constant has not changed down to a monolayer thickness. If the films are prepared by evaporation, the situation is different, since they are not continuous and their structure is porous (see Chapter 4) so that the dielectric constant decreases. The threshold thickness below which the constant starts to change depends on the structure, i.e. on the actual deposition conditions such as material and temperature of the substrate, deposition rate, etc. An example of such a dependence is given in Fig. 126. On the other hand amorphous

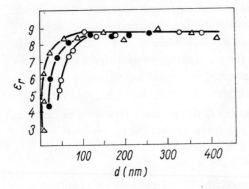

Fig. 126: The dependence of dielectric constant $\varepsilon_r$ of ZnS on film thickness: $\Delta - Al - ZnS - Al$ on mica; ● — the same system on glass; ○ — Au—ZnS—Au on glass (Chopra).

oxides prepared by anodic or thermal oxidation are continuous down to thicknesses of several nm (as confirmed by measurements of the tunnel current). These films are actually utilized in capacitors, namely, $Ta_2O_5$ and $Al_2O_3$ (the anodic-oxidized ones may be used from 10 nm up, the thermally oxidized from 50 nm up). Evaporated films may be used; of SiO their properties depend largely on the deposition conditions (pressure of oxygen, deposition rate, etc.). It has recently been found that very suitable properties in this respect are exhibited by some dielectrics deposited by sputtering.

The dielectrics most thoroughly studied as far as their dielectric properties and energy structure are concerned are alkali halides. They are of no use in capacitors due to their solubility in water. Nevertheless, the dielectric properties of thin films of these materials have been studied in great detail since they can provide valuable information on mechanisms of various processes.

Dielectric losses arise if the polarization in a dielectric is unable to follow variations in the electric field. The frequency and temperature dependences of the losses provide information on relaxation processes. A comparison of the losses of bulk and thin film materials should reveal a discrepancy if the structure of the film differs from the standard structure or the thickness is so small that the surface of the film will influence the orientation of internal dipoles. As we have seen in earlier chapters, the film stoichiometry, the structure, and the number and type of defects all depend a great deal on the deposition conditions and so, consequently, will the dielectric losses.

A characteristic of thin films of alkali halides is that dielectric losses of evaporated films decrease with time after deposition, a consequence of film ageing. This is observed mainly at low frequencies as is evident from Fig. 127. The effects of ageing increase with increasing radius of the cations

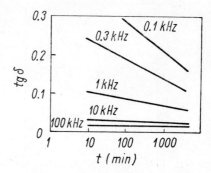

*Fig. 127:* The loss angle of NaCl as a function of time measured from the film completion (for various frequencies).

and decrease with increasing radius of the anions. Two regions of prominent losses have been observed in the loss vs. frequency plot, namely, at about 1 Hz and in the region of very low frequencies of about 0.01 Hz.

The phenomenon has been studied in great detail by Weaver, who suggested that the basic loss mechanism here is due to the migration of cation vacancies. The concentration of vacancies in these substances can be quite large $(\sim 10^{19}\ \mathrm{cm}^{-3})$. The ageing consists in the disappearance of vacancy pairs, the rate of the process being determined by the mobility of the slower anion vacancies. Since the dc conductivity of these materials is relatively low and the conduction process is accompanied by polarization effects, it was suggested by Weaver that the vacancy migration is blocked at intercrystallite boundaries. The blocking is responsible for the existence of a loss peak at a relatively high frequency. The vacancy migration is not, however, stopped completely and a certain current always flows between the electrodes; a very-low-frequency loss peak should be ascribed to this current. The peak becomes more pronounced with increasing deposition temperature, i.e. with increasing size of the crystallites. A similar effect is brought about by air moisture which also enhances the size. The activation energies corresponding to the two peaks are 1 eV and 0.7 eV for NaBr.

In view of these concepts, lower losses had been expected in mono-crystalline epitaxial films compared with those in polycrystalline films. And this has been actually observed. The measurements have been done on LiF epitaxial films and, within the frequency range 0.01 Hz to 100 kHz, the losses are at least an order of magnitude less than for polycrystalline material and a high-frequency loss peak has not been found. At low frequency the losses are comparable for both kinds of material, which is in accord with their assumed origin in polarization effects near the electrodes.

The amorphous films of oxides prepared by anodic oxidation (namely $Al_2O_3$ and $Ta_2O_5$) are remarkable for a constant tg $\delta$ in the range of audio

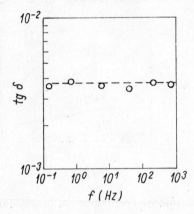

*Fig. 128:* The loss angle of $Ta_2O_5$ as a function of frequency.

frequencies. In view of their amorphous structure it is not unexpected that no pronounced peaks appear and that relaxation times are distributed continuously over a certain range. An example of the frequency dependence of tg $\delta$ for $Ta_2O_4$ is shown in Fig. 128.

In SiO films, which are frequently used as dielectrics, measurements of the dielectric loss ranging from $10^{-2}$ to $10^7$ Hz show a peak at 0.1 Hz. The corresponding activation energy is 0.4 eV. It is believed that the losses arise from the motion of free silicon ions.

The losses at low frequencies result from the ion motion, but such motion is not possible at high frequencies since the ions cannot follow the fast variations of the electric field, and thus the losses are then caused only by the electron part of the polarization. The permittivity is then in effect equal to the square of the optical refractive index.

Another significant property of dielectrics is their dielectric strength. There are several mechanisms of electrical breakdown, including thermal breakdown, resulting from a conductance regime in which a considerable part of the energy of electrons is transferred to the crystal lattice, the Joule heat is liberated and the conductivity rises till thermal ionization occurs. This mechanism leads to a destruction of the dielectric. The avalanche breakdown occurs when electrons are accelerated so much that they can excite additional electrons from the valence band of the dielectric. These, in turn, take part in further acceleration-kicking process so that the process resembles an avalanche. If this process is to be established, an electron has to acquire during its free path an energy sufficient to excite another one. This means that the film thickness must be greater than the mean free path of the electrons. Similar breakdown is that due to extensive ionization of impurities and defects, and that represented by the process mentioned in the preceding section, in which a strong field induces the tunneling of electrons from the valence to a conduction band of a dielectric. All tnese mechanisms, except the first, need not produce irreversible changes in the dieletric.

The experimental results indicate that the breakdown voltage is relatively high in thin films and corresponds to the intensity of field approaching that for the tunneling breakdown. To give an example, the value of $2.3 \times 10^7$ V/cm is reported for $Al_2O_3$ thin films. An increase in the dielectric strength with decreasing thickness has been observed in many materials, the threshold under which the increase appears varying greatly with materials (e.g. it is 20 nm for ZnS, 1600 nm for NaCl) and is usually much greater than would be expected from the known mean free paths of electrons. In spite of the great practical importance of these effects, which have inspired a number of theoretical and experimental works, their physical nature is not yet sufficiently elucidated.

### 6.2.8 Ferromagnetic Properties of Thin Films

When speaking about magnetic films, we usually mean ferromagnetic ones. Para- and diamagnetic effects in thin films are hardly observable because they are too small. On the other hand, the study of the properties of ferromagnetic thin films has contributed to the explanation of ferromagnetism.

The ferromagnetic films mostly studied have been those of ferromagnetic metals (i.e., Fe, Ni, Co, Gd) and ferromagnetic alloys. To a smaller extent ferromagnetic films of some dielectric materials, mainly ferrites, have also been investigated. As regards methods of preparation, practically all are applicable here, including evaporation, cathode sputtering, non-vacuum techniques such as electrolytical deposition and thermal decompositon, etc. Most of the work has been devoted to evaporated films. It has been found that ferromagnetic properties depend considerably on the substrate temperature, deposition rate and composition, but are not yet very sensitive to pressure variations provided no oxidation of the film occurs.

As is well known, in ferromagnetic materials regions of spontaneous magnetization exist in which the orbital and spin magnetic moments of electrons in individual atoms align themselves spontaneously along a certain direction due to their mutual interaction. These regions, called domains, are randomly oriented under normal conditions so that the material does not outwardly exhibit any magnetic moment. On applying an external magnetic field, the domains whose orientation is close to that of the field begin to grow and other domains are rotated so that at a certain intensity of the magnetic field saturation magnetization is reached, when all domains have the same orientation. Further increase in the field intensity produces rotation of the direction of magnetization so as to approach the direction of the external magnetic field.

In bulk ferromagnetic materials in single-crystal form there are certain directions of easy magnetization. For example, in Ni the easiest direction is the (111), followed by the (110) and the (100). The order is reversed in Fe. In the unmagnetized state, the domains are oriented along these directions. The origin of the occurrence of such directions is in the interaction of the orbital moment and spin of electrons with the periodic electric field of the crystal lattice. These properties are preserved in ferromagnetic mono-crystalline substances down to very small thicknesses, provided the films are prepared in the absence of an external magnetic field. The action of an external field gives rise to a considerable anisotropy which depends on the field direction. The anisotropy has been studied in an alloy of Ni and Fe (so-called permalloy: 81% Ni, 19% Fe) which possesses a very low coercivity and is important in practical applications.

One of the fundamental questions concerning ferromagnetic thin films is how the basic quantities such as magnetization $M$ and the Curie point (i.e. the temperature above which the material loses its ferromagnetic character and becomes a paramagnetic one) depend on the thickness of a film. There are two theories of the ferromagnetic state: the molecular field model and the spin-wave approach. Both theories predict retainment of ferromagnetic properties down to thicknesses of the order of $2-3$ nm. The experimental verification of these conclusions is difficult and at the same time problematic, since the theories consider ideally parallel-sided layers whereas films of such a small thickness have crystallite structure and in effect the thickness value only denotes the mean thickness. The best hopes for success in this respect are in work with epitaxial films. Very good agreement with theory has been achieved in measurements on epitaxial films of Ni and Fe-Ni on copper deposited in turn in the form of an epitaxial layer on mica (Fig. 129).

The occurrence of magnetic anisotropy indicates that the free energy of a ferromagnetic substance is a function of the direction of magnetization $M$. The easy directions of magnetization correspond to the minima of the energy and the difficult directions of magnetization to the maxima of the energy. In the case of uniaxial magnetic anisotropy (when there is one easy direction of magnetization), which is the most important one for thin films, and provided $M$ lies in the plane of a film (which in practice is a condition fulfilled in thin films), the energy E is given as

$$E = K_A \sin^2 \vartheta \qquad (6.106)$$

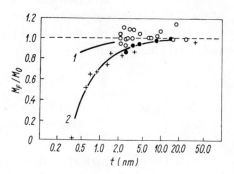

*Fig. 129:* Dependence of magnetic moment $M_f$ on film thickness $t$ ($M_f$ is related to the bulk value $M_0$): (1) molecular field theory result for Ni(111); (2) spin-wave theory result for Ni(111); ● — Stünkel, ○ — Neugebauer, + — Gradmann and Müller.

where $K_A$ is the constant of uniaxial anisotropy and $\vartheta$ is the angle between $M$ and the easy direction of magnetization (ED). Thus two states of minimal energy exist, for $\vartheta = 0$ and $\vartheta = \pi$, i.e. the equilibrium states are those with $M$ parallel or antiparallel to ED. The energy maxima occur for $\vartheta = \pi/2$ and $\vartheta = \frac{3}{2}\pi$. It is important in practical application that an external field acting in the plane of a film at an angle to the direction of ED may effect a switch of magnetization from one stable orientation to another one. The total energy during operation of the external field is given as the sum of the energy (6.106) and the energy of interaction with the external field, i.e.

$$E = K_A \sin^2 \vartheta - H_e M \cos \vartheta - H_h M \sin \vartheta$$

$$\left(H_e = H \cos \beta, \quad H_h = H \sin \beta\right) \qquad (6.107)$$

where $H_e$ and $H_h$ are the components of the external field along the easy and hard axes respectively. The conditions for an energy minimum are

$$\frac{dE}{d\vartheta} = 0 \qquad (6.108)$$

$$\frac{d^2E}{d\vartheta^2} > 0 \qquad (6.109)$$

which lead to the expressions

$$H_A \sin \vartheta \cos \vartheta + H_e \sin \vartheta - H_h \cos \vartheta = 0 \qquad (6.110)$$

$$H_A(\cos^2 \vartheta - \sin^2 \vartheta) + H_e \cos \vartheta + H_h \sin \vartheta > 0 \qquad (6.111)$$

The symbol $H_A$ denotes an anisotropy field defined as

$$H_A = \frac{2K_A}{M} \tag{6.112}$$

From (6.110) the hysteresis loop may be calculated for any arbitrary direction of the field. Let us consider two extreme cases:

(1) The field is parallel to the ED, i.e. $\beta = 0$. Then by solving (6.110) and (6.111) for $H_h = 0$, we get

$$H_e = - H_A \cos \vartheta \tag{6.113}$$

with $\vartheta$ being either 0 or $\pi$. Thus the transitions between $\vartheta = 0$ and $\vartheta = \pi$ occur at $H = \pm H_A$. The hysteresis loop has the rectangular shape shown in Fig. 130a.

(2) The field is parallel to the hard direction (HD). In this case $H_e = 0$ and in the same way we obtain $\vartheta = \pm \pi/2$. The pertinent hysteresis loop is given in Fig. 130b. The loop indicates that the magnetization rotates with

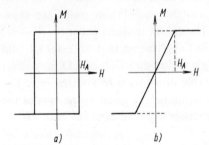

a)    b)

Fig. 130: Hysteresis loops: (a) field in easy axis direction; (b) field in hard axis direction.

increasing field from its initial orientation ED into the HD orientation. After the field exceeds $H_A$ the magnetization is competely parallel and changes no more. Hysteresis loops for other directions of the external field may be similarly determined, their shapes being then crossbreeds of the two cases discussed above.

In general, hysteresis loops found experimentally agree with the predictions quite well, but there are some differences in detail. In particular, the upright parts of the rectangular hysteresis loop are in fact not infinitely steep and the coercivity is thus smaller than $H_A$ and not equal to it (as the theory would have).

In the case illustrated in Fig. 130a, there exist two stable states with magnetization parallel and antiparallel to the field direction. Such a system may serve as a memory element in digital computers (it memorizes the digit 0 or 1 in the binary system). If the element is to operate rapidly, it is necessary that the switching from one stable state to another one should occur in a very short time. This requirement is met by magnetic thin films since the switching time for them is of the order of nanoseconds. Typical parameters for polycrystalline permalloy films are $H_A \approx 3\,\mathrm{Oe}$, coercivity $\sim 1\,\mathrm{Oe}$; a field of a few oersted is adequate for the switching.

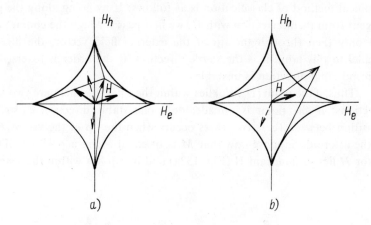

$a)$ $\qquad\qquad\qquad\qquad\qquad$ $b)$

*Fig. 131:* Stoner-Wohlfarth asteroid.

The conditions for switching can be determined from the commutation asteroid, which is constructed in the following way: The ED and HD axes are chosen as the coordinate axes and (6.110) is considered as representing the equation of the line which corresponds to an angle $\vartheta$. The envelope of all such lines is given by

$$H_A(\cos^2 \vartheta - \sin^2 \vartheta) + H_e \cos \vartheta + H_h \sin \vartheta = 0 \qquad (6.114)$$

The envelope may be expressed alternatively as

$$H_e = -H_A \cos^3 \vartheta \qquad (6.115a)$$

$$H_h = H_A \sin^3 \vartheta \qquad (6.115b)$$

By eliminating $\vartheta$ from these two relations we obtain the equation of the asteroid in the form

$$H_e^{2/3} + H_h^{2/3} = H_A^{2/3} \qquad (6.116)$$

which is illustrated in Fig. 131. The asteroid is used as follows: Through the point $(H_e, H_h)$, which represents the applied external field, tangents are drawn to all branches of the asteroid. In the case $(a)$, when the point lies within the figure, there are four such tangents; in the case $(b)$, when the point lies outside the asteroid, two tangents exist. The equilibrium directions of magnetization are parallel to those of the tangents. Two of them correspond to stable equilibrium (drawn in boldface), the other two to unstable equilibrium (dashed arrows). If the contact points lie in quadrants I, II, III or IV, then the tip of $M$ lies in quadrants II, I, IV, or III, respectively. The stable equilibria are those for which the inequality (6.111) holds. The graphical method of classification is as follows: If by going along the given tangent from its intersection with $H_e$ we first pass through the contact point and only then through the tip of the external field vector, the direction parallel to this tangent is the stable direction. If the order is reversed, the corresponding direction is unstable.

Thus, if the point $(H_e, H_h)$ lies within the asteroid, there are two stable directions; if the point is outside, only one stable direction exists. The transition between these two cases occurs when the tip of the vector $H$ lies on the asteroid. Suppose now that $M$ is oriented along a positive ED, the vector $H$ lies in quadrant II (Fig. 132a) and its tip lies within the asteroid.

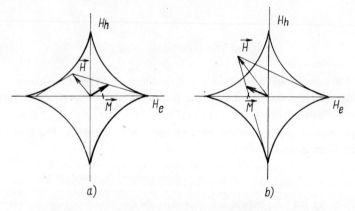

*Fig. 132:* Diagram illustrating the graphical determination of the field needed for magnetization reversal.

On applying this field, the magnetization vector is rotated into the direction $M$ indicated in the diagram. After the field is removed the magnetization returns to its initial position. If the field exceeds the threshold value determined by the asteroid, then the magnetization is reversed to the direction of $M$ indicated in Fig. 132b and finally, upon removal of the field, assumes the direction opposite to the initial one. The magnetization can be switched back again into its original position by applying a field whose vector lies in the first quadrant and whose magnitude again exceeds the threshold one.

This model, called the Stoner-Wohlfarth model, describes the behavior of uniaxially anisotropic ferromagnetic films satisfactorily and the concepts discussed can be generalized to cover systems with biaxial anisotropy or with combinations of uniaxial and biaxial anisotropy.

However, deviations are often observed from the behavior predicted by this model. One explanation of the inadequacy of the model is the fact that the direction of magnetization is not the same throughout the whole specimen as there are quasiperiodic deviations from the mean direction through a range of several degrees (Fig. 133). The deviations can be made visible by the electron-microscopic method of Lorentz (see later). Their origin is seen in local magnetic inhomogeneities, which in turn are ascribed to the occurrence of randomly oriented crystallites in the film.

*Fig. 133:* Scheme of magnetic ripple.

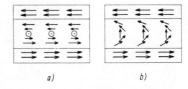

*Fig. 134:* Domain walls: (a) Bloch; (b) Néel.

As has been noted already, ferromagnetic materials contain certain regions the domains of which are oriented in certain direction due to spontaneous magnetization. Such domains generally occur also in thin ferromagnetic films. On passing through the boundary between domains the spin directions do not change abruptly; there is a transition region in which the spins change gradually. If this region, called a domain wall, is thick, then the directions of adjacent spins differ little and consequently the spin-spin interaction energy is small; however, the volume by which the magnetization orientation differs from ED is large. On the other hand, if the wall is thin, the transition volume is small but only at the price of a large spin-spin interaction energy.

In thin films the magnetization lies practically in the plane of the film since any departure from this direction would give rise to a large demagnetizing field. In films produced in a magnetic field, field-induced anisotropy dominates. Let us consider the most important case of two adjacent domains of opposite orientation. The spin directions can change either in the manner shown in Fig. 134a or in that shown in Fig. 134b. The former case, when the spins rotate about an axis normal to the wall, is frequent in bulk ferromagnetic materials and is called the Bloch wall. Such arrangement in a thin film would lead to a large demagnetization field at the wall center. In this case the spins rotate in the plane of the film and the resulting transition region is called the Néel wall. These walls are found in very thin films, e.g. for permalloy, up to about 30 nm. The Bloch walls dominate in films more than 90 nm in thickness. In the intermediate-thickness region, a special type of wall is formed, called a cross-tie wall. Their formation and structure is shown in Fig. 135. These walls are in effect Néel walls in which there are periodically spaced regions of spins with an opposite sense of rotation.

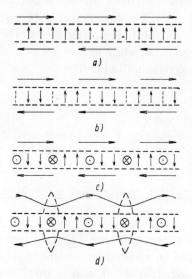

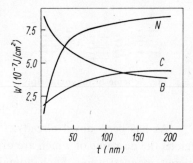

*Fig. 135:* Schematic illustration of individual stape in the formation of a "cross-tie" wall.

*Fig. 136:* Theoretical curves for the energy density of a domain wall as a function of thickness: B — Bloch wall; N — Néel wall; C — cross-tie wall.

Between the regions, there are very narrow zones (about 10 nm) in which the magnetization is perpendicular to the film. These are the Bloch lines. The dependence of the wall energy on film thickness is shown in Fig. 136 for all the three wall types.

Finally, we shall briefly mention some methods for examination of the magnetic properties of thin films. A method very much in use which allows details of the magnetic structure to be made visible is the Bitter technique. In this a colloidal suspension of fine ferromagnetic particles $(Fe_3O_4)$ is spread over the specimen. Its surface is then covered by a glass cover slip and examined in a light microscope (often in the dark-field mode). The regions with higher magnetic fields will attract more particles and domain walls will become visible by exhibiting markedly higher aggregations of particles in contrast to other areas. The technique is simple but it cannot detect finer details of the structure.

A very sensitive method is Lorentz electron microscopy. The parameters of an electron transmission microscope are adjusted so that the objective lens focus is at a plane several centimetres from the film. Owing to the action of the Lorentz force $e(V \times B)$, the trajectories of electrons are deflected and thus the intensity of the electron beam is increased or decreased at a given site of the image plane, depending on the local direction and intensity of the magnetic field. The resolution of this method is considerable so that it can be used for the observation of magnetization ripple. Under certain conditions it may be possible to assess the direction and intensity of the film magnetization at individual sites.

## 6.3 Optical Properties of Thin Films

Some of the most important aspects of thin film optics have been discussed in Sect. 3.3 in connection with optical thickness-measurement techniques. But some additional remarks will be added here.

The fundamental physical properties investigated by thin film optics are reflectance, transmittance and polarization of light for films at various wavelengths and angles of incidence of the beam. These properties are

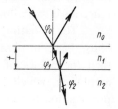

*Fig. 137:* Diagram of light path on transmission through a thin film.

determined from the electromagnetic theory of light as a function of the complex refractive index and thickness of the film. The reflectance and transmittance obey the Fresnel relations. If $n_0$ is the index of the medium from which the light is incident, $n_1$ the film index and $n_2$ the index of the medium into which the light enters from the film, then the reflectance $r_1$ $(r_2)$ and the transmittance $t_1$ $(t_2)$ (Fig. 137) at the $n_0/n_1$ interface (or the $n_1/n_2$ interface) are given for nonabsorbing film as

$$r_{1\,\mathrm{II}} = \frac{n_1 \cos \varphi_0 - n_0 \cos \varphi_1}{n_1 \cos \varphi_0 + n_0 \cos \varphi_1} \tag{6.117}$$

$$t_{1\,\mathrm{II}} = \frac{2n_1 \cos \varphi_1}{n_1 \cos \varphi_0 + n_0 \cos \varphi_1} \tag{6.118}$$

for light whose electric vector lies parallel to the plane of incidence, and

$$r_{1\perp} = \frac{n_1 \cos \varphi_1 - n_0 \cos \varphi_0}{n_1 \cos \varphi_1 + n_0 \cos \varphi_0} \qquad (6.119)$$

$$t_{1\perp} = \frac{2n_1 \cos \varphi_1}{n_1 \cos \varphi_1 + n_0 \cos \varphi_0} \qquad (6.120)$$

for light polarized with the electric vector perpendicular to the plane of incidence. The resultant reflectance and transmittance are then determined by summing multiply reflected and transmitted beams, taking account of the phase differences $\left(\text{e.g. } \delta_1 = 2\pi v n_1 t \cos \phi, \text{ where } v \text{ is the wavenumber of light and } t \text{ is the thickness}\right)$. For isotropic film the sum may be written as

$$R = \frac{r_1^2 + 2r_1 r_2 \cos 2\delta_1 + r_2^2}{1 + 2r_1 r_2 \cos 2\delta_1 + r_1^2 r_2^2} \qquad (6.121)$$

$$T = \frac{n_2 t_1^2 t_2^2}{n_0 \left(1 + 2r_1 r_2 \cos 2\delta_1 + r_1^2 r_2^2\right)} \qquad (6.122)$$

Equations (6.117) to (6.120) give the Fresnel coefficients for the parallel and perpendicular components of the reflected $(r)$ or transmitted $(t)$ light. The reflectance variation with film thickness is shown in Fig. 31.

The situation is much more complicated in the case of absorbing films. The film parameters are only measured at normal incidence because then the planes of constant phase lie parallel to the planes of constant amplitude, which is not generally the case for oblique incidence. Light propagation is now described by a complex index $\left(n_1 - ik_1\right)$, and the form of the wave is given by the expression

$$\exp\left(-\frac{2\pi k_1}{\lambda_v}\right) \exp 2\pi i v \left(\tau - \frac{n_1 z}{c}\right) \qquad (6.123)$$

where $\lambda_v$ is the vacuum wavelength of the radiation, $\tau$ the time, $z$ the propagation coordinate, and $c$ the propagation velocity.

Expressions for reflectance and transmittance are obtained by substituting the index $\left(n_1 - ik_1\right)$ in the Fresnel relations $(6.117)-(6.120)$.

In strongly absorbing materials $\left(\text{as are, e.g., metals in the visible region}\right)$ the transmittance decreases with increasing thickness and does not exhibit oscillations similar to those exhibited in Fig. 31. In metals the transmittance is only about 1% for 100-nm films.

The refractive index of a thin film is in some cases equal to that of the bulk material. In other cases, the index is lower $\left(\text{e.g. in CaF}_2\right)$, which is due to a porous structure. In such cases the porosity may be determined from the index.

A very important field of thin film optics is that dealing with the multi-layer systems. By combining various numbers of films of certain thicknesses and refractive indices, it is possible to achieve a preselected wavelength dependence of reflectance or transmittance. In this way, it is possible by combining dielectric films (e.g. $CeO_2$ and $MgF_2$, 15 films in total) to attain a reflectivity approaching 100% for certain $\lambda$ (the theory yields 0.9996; in practice it is well above 0.99).

In a similar manner, antireflection coatings or filters with bands of transmission of various widths can be formed. In the latter, even extra-ordinarily narrow bandwidths (of 1 nm) have been obtained.

Calculation of more complex optical systems with preselected param-eters is a very laborious task which is nowadays carried out with the help of digital computers.

CHAPTER 7

# APPLICATION OF THIN FILMS

There are already so many applications of thin films that only a small number of them can be mentioned here.

As has been shown in the foregoing chapters, thin films exhibit remarkable optical, electrical and other properties. In addition a thin film may considerably influence various processes occurring at a surface or interface, such as corrosion, friction, etc. In sliding bearings, for example, evaporated iridium thin films are used on account of their lubricating power. Use can also be made of other elements with high ductility such as lead or selenium. The surface resistance of glass may be enhanced by depositing a layer of $Al_2O_3$ over its surface (this naturally would affect the optical properties substantially), in other cases the glass is covered by a layer of SiO which modifies the surface tension and the possibility of chemical reactions at the interface. Relay contacts are coated with thin films of rare metals in order to prevent burning, etc.

In electron microscopy, use is made of oblique deposition of a thin film of some heavy metal onto the specimen under examination with the result that the shadowed surface so produced reveals its profile (see Sect. 2.22).

The bulk of thin film applications lies, of course, in the fields of optics and electronics.

## 7.1 Optical Applications

The optical applications of thin films were the first which came into widespread industrial use. In 1912, mirrors were first made by evaporation of metals. From that date, thin films of absorbing substances have been used extensively for optical apparatus. Astronomical reflectors may serve as an example. For the reflecting coatings of mirrors, aluminium, rhodium and sometimes silver are used. Rhodium has 80% reflectance in the visible region and displays great mechanical strength and chemical resistance.

Aluminium possesses still greater reflectance, about 90%, but is not stable chemically and thus its properties remain constant only in clean surroundings. Silver is soft and is modified chemically mainly in the presence of sulfur compounds. At wavelength 550 nm, its reflectance is 97%, in near-ultraviolet region the reflectance decreases so that at 320 nm it is only 8%. The coating thickness is chosen so as to allow about $10^{-3}\%$ of light to be transmitted through the mirror. Thicker coatings are not advisable since they lead to a greater grain and increased dispersion.

Besides their use for mirrors, semi-transparent reflection coatings are used in a number of optical devices. For this purpose platinum, rhodium and chromium are utilized. The reflectance of these materials remains constant throughout the whole visible region even at various film thicknesses so that they may be utilized for attenuation of visible light without changing its spectral composition, i.e. for formation of gray filters. The transmittance may be monitored optically during the production. In some cases (e.g. in chromium) it is necessary to take account of the fact that the transmittance will later change owing to surface oxidation.

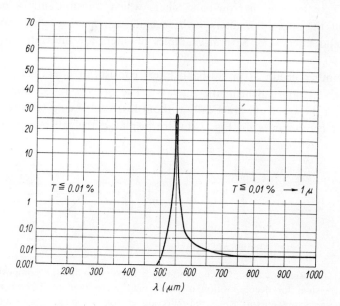

*Fig. 138:* Charactristic of monochromatic filter Filtroflex-B20.

Thin films of gold are used on account of their selective reflectance and transmittance for reflecting heat radiation.

Still wider and more diversified uses in optics have been found for thin films of non-absorbing materials by utilizing the interference phenomena

226

(see Sect. 3.3.2 and Sect. 6.3). By combining films of suitable thicknesses and refractive indices, antireflection coatings can be produced on glass surfaces which enhance the transmittance of the system. This is important especially in optical systems with many glass-air interfaces (e.g. photographic objectives).

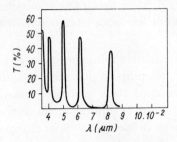

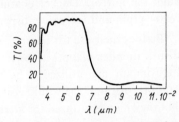

*Fig. 139:* Characteristic of polychromatic interference filter.

*Fig. 140:* Characteristic of interference filter for reflection of heat radiation.

In a similar way it is possible to produce interference filters of various types. These includes narrow-band filters which transmit only a narrow region of the spectrum, filters with several such narrow-band windows, broad-band filters, etc. Several examples of filters are illustrated in Figs. 138 to 140.

Another group of broad-band filters consists of filters used for the selection of colors (dichromatic mirrors). The mirrors reflect some components of the spectrum practically without loss while being poor reflectors for other components. In Fig. 141a are displayed the reflectances of two

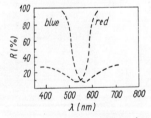

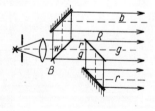

*Fig. 141:* (a) Reflectivities of two interference mirrors used for decomposition of light; (b) system for decomposition of white light using two interference mirrors: B — blue-reflecting mirror; R — red-reflecting mirror; W, b, g, r — white, blue, green and red light, respectively.

such filters, one reflecting practically all blue, another all red. By using the mirrors white light may be decomposed into three colors as is schematically shown in Fig. 141b. This decomposition is used in the formation of colored images in film or TV.

The third group of materials utilized in thin film optics is formed by weakly absorbing substances (i.e. those in which the light intensity decreases to $1/e$ only after passing over a path longer than the wavelength of the light used). The transmittance in these materials usually increases with $\lambda$, i.e. they display a colored hue in transmission, mostly yellow or brown. They are used for reduction of light intensity, (e.g. in sun-glasses). These materials, which are mostly inorganic, have thus no selective absorption (e.g. $SiO$, $MgF_2$, $Na_3Al_6$, etc.). On the other hand organic materials often possess selective absorption, in which case they are used as color filters. Weakly absorbing materials usually have a high refractive index and a high reflectance. The effects of these undesirable properties may be overcome by combining the materials with a layer of non-absorbing substance with suitably low index.

A combination of metal and non-absorbing films is also utilized in the protective coating of aluminium mirrors by an $SiO$ layer as mentioned above. Such a coating introduces, however, a certain degree of spectral selectivity into the reflectance which may be ascribed to interference phenomena. By combining metal together with non-absorbing films ultranarrow-band filters can be produced based on the Fabry-Perot interferometer principle, with the difference that a dielectric film is inserted between the semitransparent mirrors instead of an air gap.

# 7.2 Applications in Electronics

The applications of thin films in electronics have been steadily growing in importance during the last ten years or so. We noted in the introduction, that a demand exists for further development of minimization of dimensions and weight of electronic systems, while stress is also laid on long-term reliability and low production costs. The construction of digital computers and that of other intricate electronic systems for measurement or control require a multiple duplication of identical functional elements (e.g. flip-flop circuits, amplifiers). This development aims at grouping these elements into independent units — integrated circuits for the production of which thin films are of paramount importance.

Individual components of an electronic system may be divided into two groups: passive ones which only transmit energy, consuming a part of it, and active ones such as vacuum tubes and transistors. The fabrication of the passive elements on a thin film base is now an industrial commonplace, whereas the development of the active elements still remains, with a few exceptions, at the laboratory stage.

### 7.2.1 Electric Contacts, Connections and Resistors

The microminiaturization of electronic circuits makes severe requirements on contacts and electric connections between individual components. Vacuum-evaporated metal films have proved to be suitable for this purpose. They can be molded into desired patterns by means of suitable masks, photolithographic methods or by cutting with electron or laser beams. These techniques have already been discussed. As for construction materials, usually non-corroding metals are used, e.g. gold. The choice of contacts is limited by their resistance characteristics.

The following requirements are made on resistors in electronic systems: accuracy of resistance value, small temperature coefficient, stability under operating conditions, low noise level and small voltage coefficient. The most perfect resistors are represented by wire-wound resistors. These, however, are expensive and defy excessive miniaturization. For comparison, some of their properties are given in Table 15 together with those of carbon resistors (produced by pyrolytic decomposition of carbon compounds) and those of thin metal films.

The Most Important Properties of Electrical Resistors                           *Table 15*

| Resistor type | Tolerance for resistance values (%) | Resistance range | Noise ($\mu$V/V) | Voltage coefficient |
|---|---|---|---|---|
| carbon film | 0.25 | $1\Omega - 50$ M$\Omega$ | $0.3 - 0.5$ | $2 \times 10^{-3}$ |
| metal film | 0.1 | $1\Omega - 10$ M$\Omega$ | 0.1 | $10^{-4}$ |
| wound wire | 0.01 | $1\Omega - 0.05$ M$\Omega$ | only hot-wire | 0 |

If evaporated resistors are made from metals with resistivity of the order of $10^{-5}$ $\Omega$ cm, high resistance values can only be achieved by formation of very thin films. Such extremely thin films are, of course, no longer continuous and this fact results in certain specific properties, which have been discussed in more detail in Sect. 6.2.2.

The resistance of thin films is usually characterized by a quantity $R_\square$, i.e. the resistance of a square film measured between two opposing sides ('sheet resistance'). It may be readily inferred that $R = \varrho/t$, where $\varrho$ is the resistivity, $t$ the film thickness ; the unit of this resistance is denoted by $\Omega/\square$. For the alloy of Ni and Cr (nichrome) the ratio is $80:20$ and $R = 500$ $\Omega/\square$ for film of 50 nm thickness.

The nichrome is evaporated from a tungsten boat either by the flash technique or by electron bombardment.

Tantalum, titanium and niobium are also used. They are deposited mainly by cathode sputtering so as to possess resistance lower than the one required. Subsequently anodic oxidation is applied to achieve the right value. The surface oxide also serves as a protective layer.

In order to increase total resistance, the films are deposited as spirals (on rods) or meanders (on flat plates). The pattern required is produced by use of suitable masks or by mechanical or electron-beam cutting.

For high values of resistance, use is made of cermet films, which are mixtures of metals and dielectrics. Their resistivity increases with the dielectric content and can be varied over a very wide range. For Cr—SiO cermet the resistivity can be varied from $10^{-5}$ to $10^1$ $\Omega$ cm, which means that for 100 nm thick film $R_\square$ can vary from 1 $\Omega/\square$ to 1 $M\Omega/\square$. The films are usually deposited on ground porcelain tubes or plates.

The stability of evaporated resistors is improved by artificial ageing, which consists in keeping the film at $150-300$ °C for several hours. Resistance decreases exponentially with time during ageing a consequence of the continuous decrease in the number of lattice defects. After a certain time the resistance theoretically reaches stability, but, in fact, the resistance varies even then due to surface oxidation, absorption of gas from the ambient atmosphere, etc. The resistors are therefore protected either by an evaporated SiO layer or by enclosing them in protective cases. Stability is especially imperilled by high currents, by electro-transport (electrical corrosion), which involves the disappearance of material near the cathode (the film thins down at some places so that holes are created) and its transport to the anode.

The properties of some thin film resistors are set out in Table 16.

Properties of Some Thin Film Resistors                                    *Table 16*

| Composition | Resistivity ($\mu\Omega$ cm) | TCR* ($10^{-6}/$°C) | Temperature range of TCR (°C) |
|---|---|---|---|
| Pd, Ag | 38 | 50 | 0—100 |
| Cu(83), Mn(13), Mg(4) | 48 | 10 | 15—35 |
| Ni(80), Cr(20) | 110 | 85 | 55—100 |
| Ni(76), Cr(20), Al(2), Fe(2) | 133 | 5 | 65—250 |
| Ta($\alpha$) | 25—50 | 500—1 800 | |

* TCR = Temperature coefficient of resistance

### 7.2.2 Capacitors and Inductances

The formation of thin film capacitors represents a very promising field since a high capacity can apparently be achieved by use of ultrathin layers of dielectric, especially when use is made of a high-permittivity dielectric (e.g. $TiO_2$, see Table 17).

Dielectric Constants for Some Materials                    *Table 17*

| Material | SiO | $SiO_2$ | $Al_2O_3$ | $Ta_2O_5$ | $TiO_2$ |
|----------|-----|---------|-----------|-----------|---------|
| $\varepsilon_r$ | 5—7 | 4 | 7 | 15—25 | 40—170 |

Capacitors with evaporated electrodes in which the dielectric took the form of paper or various foils, were produced in the past. These capacitors have one great merit, namely, that a breakdown at a given site leads also to the evaporation of the ambient area of the electrode so that the capacitor does not stay short-circuited; the damaged region is self-healing.

However, the formation of microminiature and integrated circuits necessitates vacuum evaporation of all three basic layers of the capacitor. The main problem is then deposition of a dielectric layer without pinholes and other defects which cause faults. They are present especially in very thin films. The usual cause is dust on the substrate and evaporation of bigger particles from the vapor source.

In this respect, better quality is achieved with dielectric layers prepared by anodic oxidation, which exhibit very uniform thickness.

In addition to its capacitance, a capacitor should have a high breakdown voltage, small loss angle and good thermal endurance. The breakdown voltage is usually of the order of $10^6$ V/cm, and the greatest values are achieved with $Ta_2O_5$ ($5 \cdot 10^6$ V/cm). Tantalum has, however, relatively high resistivity, so that its use leads to an increase in loss angle (by increasing series resistance). This intrinsic increase is countervailed by deposition of a good-conducting layer, e.g. gold or platinum on tantalum electrodes.

During the evaporation of dielectric compounds partial dissociation of the material usually takes place, which produces an increase in loss angle. The materials most frequently used are: the oxides of silicon, $Al_2O_3$, and $Ta_2O_5$. Specially adapted sources have to be used for evaporation of SiO to prevent spitting of bigger particles, which create defects in the deposited film and which are responsible for faults. Of late, growing use has been made of electron-bombardment evaporation and, to a greater extent of cathode sputtering.

It is true that $SiO_2$ and $Al_2O_3$ do not possess high permittivity (see Table 17), but they exhibit good thermal endurance (e.g. $Al_2O_3$ can be used up to 400 °C). With $Ta_2O_5$, a capacity up to 0.22 $\mu F/cm^2$ for a voltage of 50 V can be attained. Higher capacitance values can be achieved by using $TiO_2$ or titanates, but these materials have so far defied all efforts to prepare them in a thin film form of adequate quality.

Concerning inductors, it is difficult at low frequencies to achieve sufficiently high values of inductance. Inductors are made in the form of a spiral which is deposited on a paramagnetic substrate (the spirals are usually of angular shape since they are deposited through masks which are readily made to shape). Higher inductance values can be achieved by depositing the inductor on a ferrite substrate. Additional improvement is effected by complementary ferrite overlay. All these treatments, however, raise production costs. These problems disappear, of course, when the inductor is intended for use at high frequencies.

### 7.2.3 Applications of Ferromagnetic and Superconducting Films

As we have seen in Sects. 6.2.5 and 6.2.8, both anisotropic ferromagnetic and superconducting films provide the possibility of information storage (since both may be transformed with relative ease from one state into another). They can thus be utilized in memory or logical units, especially in those

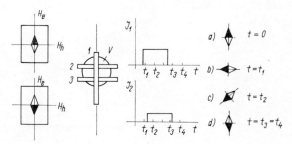

*Fig. 142:* Process of magnetization reversal in ferromagnetic memory element.

of digital computers. Anisotropic magnetic films are made for this purpose by evaporation of permalloy from a ceramic crucible onto a glass or ceramic substrate maintained at 300 to 400 °C in a magnetic field of $\sim 100$ Oe. In this way the direction of easy magnetization is established.

Reorientation of magnetization takes place either through rotation of the vector **M** in the plane of a film or by direct reversal. The latter process is relatively slow ($\sim 1$ μs) when compared with the former ($\leq 1$ ns). Thus it is desirable to effect reorientation through the rotation. The corresponding element is shown in Fig. 142. Across a ferromagnetic film three conductors (mutually insulated and also from the substrate) are evaporated. Conductors 1 and 3 are the recording lines, 2 is the reading line. At the beginning the film magnetization is oriented along one of the easy directions (indicated by the magnetic needle in Fig. 142a). By a current pulse $I_1$, the magnetization is rotated into the hard direction (Fig. 142b); the pulse $I_2$ brings about the position in Fig. 142c. After both pulses the magnetization goes spontaneously into the nearest easy direction (Fig. 142d). The reading, i.e. the process of finding the magnetization state (one of the orientations stands for 0, the other one for 1) is carried out by sending a pulse into the reading line. The pulse induces a current in line 3, the direction of which depends on the magnetization of the film.

The operation speed of the memory element is not normally determined by the speed of the reorientation but by the speed of transition processes in connected electrical circuits. Besides this planar arrangement the memory element can be made in the form of a cylinder (Fig. 143.) Wires (usually made from CuBe alloy) are coated electrolytically with a ferromagnetic permalloy layer. The axis of magnetization lies on the periphery of the layer and is established by action of a galvanizing current. The writing line $P$ runs perpendicularly to the axis, as can be seen in the figure. The relative advantage of this configuration lies in the fact that magnetic circuits of individual elements (bits) are closed, i.e. there are no demagnetizing fields.

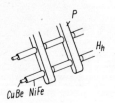

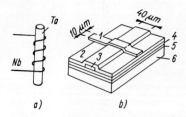

*Fig. 143:* Schematic drawing of ferromagnetic memory element in cylindrical configuration.

*Fig. 144:* Cryotron: (a) wire; (b) thin film; 1 — 0.6 μm Pb; 2 — 1 μm SiO; 3 — 0.9 μm Sn; 4 — SiO; 5 — Pb; 6 — glass.

In magnetic memory elements, the greatest problems are met in connection with the homogeneity of the film. The easy axis has not in fact identical direction everywhere because of nonuniform thickness and internal stress. The coercivity $H_K$ which according to the theory should be equal to the saturation value $H_A$ is in practice smaller (higher, in the case of so-called inverse films). The $H_K$ depends on film thickness and usually equals 0.5 to 3 Oe; the $H_A$ depends on the substrate temperature, deposition rate and impurities in the film and is equal to 2 to 5 Oe.

The superconducting switching elements, cryotrons, possess a very high operation speed and require little space and energy. These properties

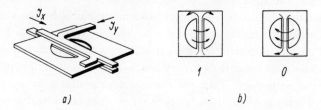

*Fig. 145:* Crowes cryotron memory element: (a) layout; (b) direction of induced current in 1 and 0 state respectively.

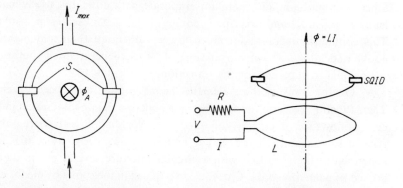

*Fig. 146:* Interference magnetometer with two Josephson junctions (S-junction).

*Fig. 147:* Clark interference voltmeter with two junctions.

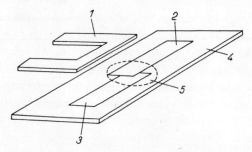

*Fig. 148:* Cryotron using Josephson effect: 1 — control loop Pb; 2 — Sn film; 3 — Sn + SnO$_x$ film; 4 — substrate; 5 — Josephson junction.

sometimes even offset the fact that they need liquid helium for their operation. The scheme of a normal cryotron (without thin film components) is shown in Fig. 144a. A niobium wire is wound around a tantalum rod. Both conductors are maintained in a superconducting state. When a current of adequate intensity passes through the wire, the superconducting state in the rod is destroyed by action of a current-induced magnetic field and the tantalum assumes normal conductance. The current must, however, be low enough not to destroy the superconductivity in the niobium wire itself. That such a situation can occur is made possible by the selection of materials used (for tantalum $T_c = 4.4$ K, whereas for niobium $T_c = 8$ K). The wire current adequate for the extinction of the superconducting state in the rod is much lower than that necessary for extinction of superconductance in the wire itself. Thus the element also operates as an amplifier. However, in this configuration the element exhibits considerable self-inductance which has detrimental effects on the transition speed.

In this respect a thin-film planar cryotron, shown in Fig. 144b, is preferable. Both conductors have the form of evaporated plane strips. Since both niobium and tantalum evaporate with difficulty, use is made of lead ($T_c = 7.2$ K) and tin. An evaporated layer of SiO serves as insulation. This configuration possesses much lower inductance than that mentioned above and allows a switch-over time of $\sim$ ns to be attained (duration of the normal to superconducting transition proper is $\sim 10^{-10}$ s).

On the basis of these elements memory units have been developed. Their function rests on the fact that in a superconducting circuit an induced current flows for practically unlimited time. We shall describe here only one type of such a unit (Crowe's). The unit is shown in Fig. 145a. It is a thin superconducting (Pb) film with a circular aperture several nm in diameter arched by a narrow spoke. Currents are induced in the spoke by means of two parallel mutually insulated control lines. The path of the current is closed via the aperture edge (Fig. 145b) and the current flows in two opposite directions, which correspond to the '0' and '1' states.

On the other surface of the film a reading line is placed parallel with both writing lines. The induced current flows through two D-shaped loops in the film. The sum of the currents induced by the writing lines is either added to or substracted from that current. In the case of substraction the spoke remains in a superconducting state and no signal is induced in the reading line ('zero'). When the currents are added together, the induced magnetic field exceeds the critical one and the spoke is transformed into the normal state.

The time constant of a circuit of normal dimensions lies in the range of 10 ns to 20 ns. Superconducting memory units appear to be very promising for the development of computers. It has been found that the consumption of helium is not excessive. For example, in Crowe's type of memory with one million bits, which would have a physical volume of 30 liters, the helium consumption would be 2 liter per hour.

On account of the negligible resistance, thin superconducting films covering the internal walls of microwave elements make it possible to realize circuits with high working current densities $(\sim 10^6 \text{ A/cm}^2)$ and resonance circuits and filters with very high $Q$-factors. To give an example: A cavity resonator has $Q = 10^9$ even at 30 GHz in spite of the fact that high-frequency surface resistance increases with frequency $(R_s \sim \omega^2)$. Thin films of Nb and Pb are used. Excellent surface quality is required for this purpose. The frequency threshold of applicability is determined by occurrence of dissociation of the Cooper pair and lies in the region of mm waves (the quanta are then comparable with the magnitude of the energy gap $\Delta(0)$). By use of such resonators oscillatory circuits with high stability can be constructed. Circuits with a reflex klystron with stability of $\Delta\omega/\omega = 3 \cdot 10^{-13}$ per 10 s and oscillators with Gunn diodes with stability of $1.3 \cdot 10^{-13}$ per 10 s were developed. It is expected that a stability of down to $10^{-15}$ per year will be attained in the future.

On account of their low noise level, similar elements are also employed in radio-communications, namely in microwave masers, parametric amplifiers, filters, low-attenuation coaxial lines, etc.

A very wide area of possible application is provided by both Josephson effects, i.e. the dc and ac ones.

Owing to strong dependence of the dc Josephson effect on the magnetic sensitivity is enhanced further by use of two parallel-connected weak junctions (see Fig. 146). The device is konwn as SQUID (Superconducting Quantum Interference Device) and is based on quantum interference between the two junctions arising from changes of magnetic flux in a loop closed by the superconductor. The critical current of the system exhibits two kinds of oscillations which are described by the relation

$$I_{\max} = I_c \left| \frac{\sin\left(\pi\Phi_s/\Phi_0\right)}{\pi\Phi_s/\Phi_0} \right| \cdot \cos\frac{\pi\Phi_A}{\Phi_0} \qquad (7.1)$$

where $\Phi_s$ is again the magnetic flux threading the junction, and $\Phi_A$ is the flux passing through the enclosed area of the loop. The sensitivity of the apparatus attains $10^{-11}$ G and may be utilized in medicine for magnetic cardio- and encephalography.

On a similar principle are based voltmeters and galvanometers which measure the magnetic field induced by a current (see Fig. 147). They achieve sensitivities of $10^{-15}$ V and $10^{-6}$ A respectively. The theoretical sensitivity limit is $10^{-19}$ V.

Systems with junctions can also be utilized in logical circuits since a weak magnetic field or electric current can switch the junction from a super-conducting into a normal state. As a result the voltage jumps from zero to $V = 2\Delta/e \sim 1$ mV in a period of about $10^{-12}$ s. The diagram of such a tunnel cryotron is given in Fig. 148. The switching time has been measured as $\sim 8 . 10^{-10}$ s, which corresponds to the rise time of the whole measuring system.

The Josephson a.c. effect can be used for microwave generation. Frequencies are in the range of 1 to 1000 GHz; the linewidth is mainly determined by the shot noise of single electrons and Cooper pairs. The spectral purity of lines is about $10^{-7}$. In junctions with point contact and series resistor $R$ the linewidth is proportional to $RT$; this fact may be utilized for temperature measurement down to a few mK.

On account of the influence of incident electromagnetic radiation upon the current-voltage characteristic of the Josephson-junction which has been discussed in Sect. 6.2.5. (see Fig. 117b), sensitive detectors of radiation can be constructed down to the submillimeter range and energy levels of $10^{-10}$ W.

Since the Josephson junction exhibits markedly nonlinear character-istics it may be used for mixing, down to the infrared region. The relation between the voltage and frequency (see equation (6.76)) together with the high accuracy with which the frequencies can be determined provide the possibility of an absolute standard of voltage assuming the precise value of $2e/h$ is known. (At the present time the constant is known to within $0.12 . . 10^{-6}$).

### 7.2.4 Active Electronic Elements

One of the most important aims of thin film electronics is the production of active thin-film elements which would substitute for the transistor.

So far there is only one such element manufactured on an industrial scale, namely Weimer's TFT (Thin Film Transistor), which utilizes the field effect. Its cross-sectional diagram is given in Fig. 149. A glass substrate supports a semiconducting film (usually CdS or CdSe) deposited between two metal strips which form ohmic contacts with the semiconductor (input and output electrode, also called source and drain electrode, respectively). The semiconductor is covered by a thin layer of a dielectric (SiO) and the structure is completed by a gate electrode. On applying a voltage between the input and output electrodes, a current determined by the resistance of the semiconductor starts to flow between the electrodes. The resistance can be modified by the bias of the gate electrode since the action of a bias-produced field affects the carrier concentration in the semiconductor layer and thus also its conductivity. If at the surface of an N-type semiconductor the energy bands are bent downwards, i.e. the electron concentration is

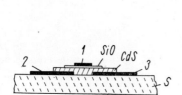

Fig. 149: Schematic cross-section of thin-film transistor (TFT); S — glass substrate; 1 — gate; 2 — input (source) electrode; 3 — output (drain) electrode

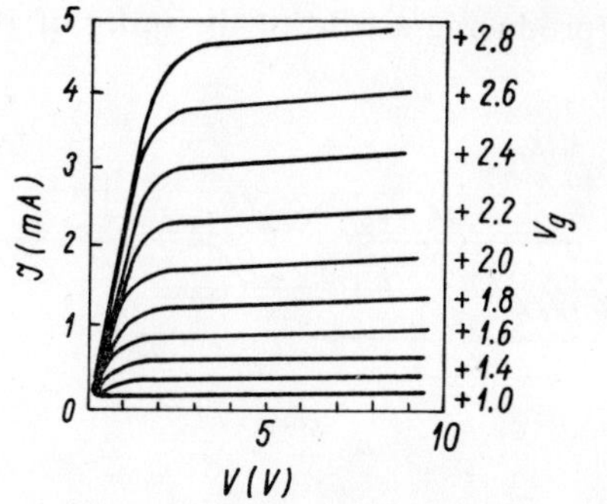

Fig. 150: Examples of characteristics of TFT (CdS n-type) in enhancement operation mode; the parameter is the gate voltage.

higher there, then a negative bias of the gate electrode causes a decrease in the current between the input and output electrodes (the depletion operation mode). If, on the contrary, the surface layer has a relatively low electron concentration due to upward bending of the energy bands, a positive bias of the gate leads to an increase in the current (the enhancement operation mode). The main parameter of the transistor is the ratio of the increment of the current between the input and output to the increment of the gate bias. In both modes saturation occurs when the conducting channel is separated from the output electrode. Examples of current-voltage characteristics are shown in Fig. 150. In order that the bias-produced effect be of sufficient magnitude, it is necessary that the number of carriers injected or removed by the field be comparable or greater than the number of carriers present in the normal state. Thus the properties depend very much on the concentration of defects (donor or acceptor levels, shallow or deep traps) in the semiconductor. The geometrical configuration is also of critical importance.

Besides the configuration shown in Fig. 149 there are a number of other arrangements, e.g. the inverse one (the gate is situated on the substrate), a planar one with two gate electrodes, etc.

In recent years experiments have been conducted with a whole array of other elements intended to serve as active components in microelectronic thin film systems. However, none of the elements has yet turned out to be ripe for practical use and we shall not therefore deal with all of them here.

As an example of such a system, the triode developed by Mead is shown in Fig. 151. It is a metal-dielectric system (of 5 nm thickness ) — the metal serves as a cathode. On application of a voltage between $K_1$ and $K_2$ (producing a field of the order of $10^6$ V/cm), electrons travel by tunneling

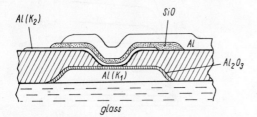

Fig. 151: Schematic cross-section of thin film triode after Mead.

from the lower negative metal layer to the upper one and the portion that does not lose too much energy in interactions with the free electrons (the thickness of $K_2$ is $\sim 30$ nm) can proceed even farther and pass through the second dielectric layer into the farther electrode.

The system therefore operates as a triode whose current can be controlled by the bias of the middle electrode. Here, the most difficult problem is that of parameter stability.

It is, however, necessary to note here that the technology of transitors in microelectronic circuits is nowadays unthinkable without thin film techniques. The technique is used in the manufacture of ohmic contacts and for deposition of metals for alloying in production of mesa-transistors and planar transistors. In recent years, transistors have been manufactured by diffusion of impurities into epitaxial films grown on a single-crystal substrate (usually Si). This technique has found a widespread use especially in the production of integrated circuits (see Sect. 7.2.6).

### 7.2.5 Microacoustic Elements Using Surface Waves

In Sect. 6.1.5 we have mentioned the possibility of utilizing surface waves in microacoustics. This area of research, which has come into existence during the last few years, appears very promising and especially so in regard to elements operating in the high-frequency region (10 MHz − 3 GHz). Low propagation velocity and small attenuation of the waves in comparison with electromagnetic ones represent their main advantages. Moreover, the systems based on surface waves exhibit a high stability and manufacturing techniques do not make any extraordinary demands on the standard method of production of microelectronic elements.

On the basis of surface waves both the passive and active components can be constructed.

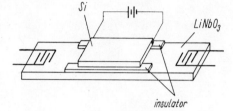

*Fig. 152:* Separated media amplifier
(circuit diagram).

As concerns the passive elements, mention has already been made of waveguides, which may also function as high-power dividers, coupling elements between two circuits, delay lines, resonators and filters. These circuits behave with distributed parameters (their dimensions are comparable to the wavelength).

There are also several types of amplifiers, among which only the separated media amplifier (SMA, Fig. 152) will be mentioned here. This amplifier seems to exhibit relatively good parameters. An acoustic wave

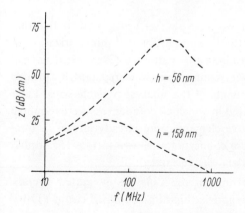

*Fig. 153:* Frequency characteristic of SMA for two different widths of air gap.

excited by an interdigital transducer propagates over the surface of a piezo-electric substance (e.g. $LiNbO_3$). The electromagnetic field thus produced penetrates the air gap into the semiconductor (Si) and acts upon electrons moving under action of d.c. voltage. The frequency response of such an amplifier is shown in Fig. 153 for two values of air gap thickness.

240

Use can also be made of the nonlinear effects to construct modulators, mixers and detectors on surface acoustic waves by means of thin films having the form of a lens or prism (Fig. 154).

The production of the elements operating on the basis of surface acoustic waves requires improvement of microelectronic techniques, espe-

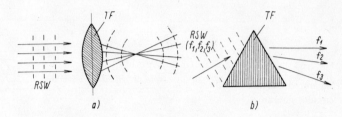

Fig. 154: (a) Focusing and (b) refraction of surface acoustic waves by means of thin film analogs of lens and prism respectively; TF — thin film; RSW — direction of incident surface acoustic wave.

cially those of surface processing and reproduction of patterns in the region under 1 μm. We refer here to electron-beam cutting, ion bombardment, deposition by molecular beams, etc.

### 7.2.6 Integrated Circuits (IC)

During the foregoing exposition we have several times pointed out that there is a universal demand for the maximum possible miniaturization of electronic systems. The first step along this path was taken when normal classical systems with soldered-wire connections were replaced by printed connections with classical components of the smallest possible sizes. Later on so-called micromodules appeared in which all components were mounted on, ceramic plates while the passive elements were already largely in thin-film form. The ceramic plates were connected as shown in Fig. 155 so that a compact block with several outlets was assembled.

Further development led from micromodules to integrated circuits. These fall essentially into two categories: thin film integrated ciruits (TFIC) and solid-state integrated circuits (SSIC or monolithic circuits). In the ideal case all components of TFIC are made by vacuum evaporation onto the insulating substrate. At present, however, this is possible only when TFT transistors are to be manufactured. Otherwise only the passive components are evaporationmade whereas active ones are fixed to the substrate as separate units. This is known as a hybrid circuit.

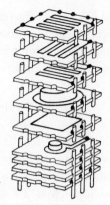

*Fig. 155:* A view of a fully assembled micromodule unit.

Another type of hybrid circuit is made on a single-crystal semi-conducting wafer (Si). Active components are made by impurity diffusion and passive ones are evaporated.

A solid-state integrated circuit consists essentially of a single-crystal of silicon in which both active and passive devices are formed by the usual techniques of semiconductor technology. The resistors are produced by doping a certain region with a suitable impurity. Insulating layers are made from silicon dioxide which is produced by thermal oxidation of the basic material. The capacitors are formed by a suitable PN-junction under reverse bias. These IC have very small dimensions. On the other hand the great temperature-dependence of parameters of all components and, in particular, the voltage-dependence of PN-junction capacitors are a drawback. Here thin films are used as contacts and connections and the production technique is similar to that of a planar transistor.

To make possible a comparison of component densities achieved in individual methods of circuit assemblage, the numbers of components in $1 \ cm^3$ will be given here: In printed wirings with miniature components, 0.4; in micromodules, 15. The volume of a memory cell in IC with bipolar transistors is only $6.25 . 10^{-4} \ mm^2$. The surface density in modern systems reaches 800 000 to 1 000 000 transistors over an area of $6 \ cm^2$.

Among the latest developments are so-called LSI circuits (Large Scale Integration). These circuits can be manufactured with several hundred gates.

Logical switching circuits nowadays make up the main bulk of micro-electronic production. In modern elements delay time is less than 1 ns (e.g. 0.7 ns). Such elements are used in ultrarapid computers with a time cycle under $10^{-8} \ s$. Thanks to the appearance of these microminiature components in recent years, the operation speed of computers has increased substantially. Memory capacities have also increased a thousand times and continue to increase. At the same time the costs involved in integrated circuits represent about $1/10$ of the costs of classical circuits, while at the same time better parameters, higher reliability and substantially smaller energy and volume requirements are achieved.

### 7.2.7 Thin Films in Optoelectronics and Integrated Optics

Devices for radiation detection form an important group of thin film semiconducting elements. They include phototubes, photo resistors, photovoltaic cells and further elements used in optoelectric and electrooptic system such as TV camera tubes, electroluminiscent panels, etc. Some of these devices will be considered here.

Photocathodes (in phototubes, photomultipliers, TV camera tubes, etc.) consist of a vacuum-evaporated thin film of metal (usually Sb or Bi) activated at high temperature in cesium or other alkali metal vapor. The resultant semiconducting compounds (e.g. alkali antimony) exhibit a photoelectric effect in the visible region. Since the work function is decreased owing to the presence on the surface of a thin film (approximately of monolayer thickness) of cesium, a certain fraction of the excited electrons can emerge into the vacuum and thereby constitute a photocurrent.

A classical representative of this group of photocathodes is the cesium antimony photocathode. The semiconducting film proper has a composition of $SbCs_3$ with a forbidden band of 1.6 eV; it is of p-type and its work function (photoelectric) is about 2 eV. The long-wavelength limit can be shifted to 850 nm through sensibilization of the surface by oxygen. As the film possesses a considerable intrinsic conductivity it may be used as a thin semi-transparent photocathode. In such cases it is deposited on a glass substrate, light is incident from the glass and electrons emerge in vacuum on the other side (the sensitivity is about $80-180\ \mu A/lm$).

Higher $\mu A/lm$ sensitivities ($\sim 250\ \mu A/lm$ and above) and spectral characteristics extending into the long-wavelength region are achieved with multialkali photocathodes which also contain other alkali elements (Na, K or Rb) besides Cs.

The latest photocathodes are formed by those with negative electron affinity. As for basic materials, use is made of compounds of the $A_{III}\ B_V$ type (as GaAs, GaP, etc.), or of Si with high concentration of p-type impurities ($\sim 10^{19}\ cm^{-3}$) in the most perfect single-crystal form available. The active surface layer is established by repeated adsorptions of ultrathin (approximately monolayer) films of Cs and O which have to take place on an atomically clean surface (ie. on a surface obtained by cleavage in ultrahigh vacuum or cleaned as much as possible by some thermal process or by ion-bombardment). The dipole moment of the surface layer produces a decrease in electron affinity (i.e. in the difference between the bottom edge of the conduction band $E_c$ and the vacuum level $E_{vac}$). At the same time there occurs a downward bending of the bands within a $\sim 5$-nm-thick region L (see Fig. 156). If an electron excited by light inside the semiconductor loses so much energy as a result of various interactions that it sinks to the bottom of the conduction band then in normal circumstances it cannot be emitted. However, owing to the aforementioned modifications it can now tunnel through the thin barrier (i.e. through the peak produced by the difference between the original and lowered affinity) and emerge into the vacuum. In fact the electron is already above the vacuum level when it is still inside the material and thus the affinity appears as if it were negative.

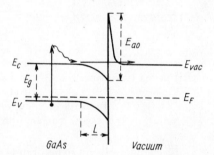

*Fig. 156:* Energy diagram of a GaAs surface with negative electron affinity.

In these photocathodes sensitivities of the order of 1 000 μA/lm are achieved and high sensitivity is also obtained in the infrared region.

Besides Cs in combination with oxygen other systems have also been used for the formation of active surface layers, namely, those in which the oxygen has been replaced by water vapor or fluorine.

The surfaces with negative electron affinities can be utilized as emitters with very high coefficients for the secondary emission (of the order of hundreds) or as cold cathodes. The secondary emitters can also be made in transmission mode provided the semirconducting film proper is thin enough for an electron beam to penetrate through it. In such a case the primary electrons impinge on the film from one side while the secondary ones depart from the other side.

A cold cathode with negative affinity can be looked upon as a pn-junction which is (a) biased in the forward direction, (b) in which the 'p' layer is thin enough for the electrons injected from the 'n' layer to be able to pass through it without being recombined and (c) which is provided with an active layer (identical with that of a photocathode). The emission mechanism is evident from Fig. 157.

244

In photoresistors, thin films of $A_{II}B_{VI}$ compounds (e.g. CdS, CdSe) with evaporated contacts are used. On application of a voltage a relatively small current, determined by small conductivity of a semiconductor, starts to flow. When the film is exposed to light, electrons are excited into the conduction band and the conductivity and the current increase substantially. In photovoltaic cells, emf is produced by exposing a pn-junction to light. Use is mostly made of Si crystals of 'n' type on which a thin film of 'p' material is fabricated by diffusion. The evaporation technique is used here to form contacts and also for deposition of an anti-reflection coating which enhances the yield.

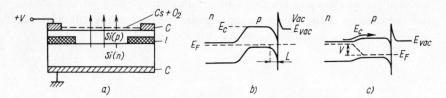

*Fig. 157:* Cold cathode with negative electron affinity:
a) construction (schematically); C — contacts, I — isolating ring (SiO$_2$), Cs + O$_2$ — activation layer
b) energy diagram without applied voltage
c) energy diagram with an applied voltage V.

Electroluminiscent panels represent a further important and promising field of thin film applications. They possess the configuration of capacitors in which, however, the dielectric is replaced by luminescent material and the upper electrode is transparent (at least for a certain region of the spectrum). An example of a typical electroluminescent material is ZnS. Application of ac voltages of 50 − 1000 Hz and of amplitudes of 150 − 500 V on the electrodes produces electroluminescence, i.e. emission of light of a certain color depending on the nature of the luminescent and electrical parameters. Such panels have hitherto been used only for signal and similar purposes but it may be reasonably expected that they will be used as light sources or as substitutes for present-day TV tubes in the not too distant future.

Recent years have witnessed a rise of a new field: integrated optics. It is, in essence, an extension of integrated circuits into the region of optical frequencies ($10^{14} − 10^{15}$ Hz, which correspond to wavelengths in the μm range). It deals with the problem of transfer of light energy by means of so-called light-guides (an analogy of normal waveguides), energy modulation (phase or amplitude modulation), demodulation, frequency filtration, generation and detection, generation of higher harmonics, modification of propagation direction, etc.

In this field, thin films can be utilized mainly in light-guides. The refractive index of the film, substrate and ambient medium together with the direction of incidence of a light wave must be chosen so that total reflection occurs at interfaces and thus the light energy is propagated inside the waveguide without any possibility of leaking before it is intended to do so. For the input and output of light energy into and from a film, special elements have to be devised. These can be prisms, gratings (prepared with the help of a photoresist on which an interference band system is projected) or wedge-shaped films. As a source, a laser is used.

### 7.2.8 Further Applications

We can easily continue the enumeration of applications. The following are of importance: thin-film resistance tensometers, mosaic or target electrodes for TV camera tubes, thin film thermocouples, etc. However, the purpose of this chapter has not been to set out an exhaustive list of all existing applications. Our intention was simply to present examples from those fields in which thin films play the greatest role and to demonstrate that their application has brought about substantial qualitative advancement. These developments are, of course, only in their early stages and further expansion and penetration of thin films into other fields is to be expected in the future.

# REFERENCES

1 Anderson, J. C. (editor): The Use of Thin Films in Physical Investigations. New York, Academic Press 1966
2 Bartl, P., Delong, A., Drahoš, V., Hrivňák, I., Rozenberg, M.: Metody elektronové mikroskopie. Praha, NČSAV 1964
3 Beam, W. R.: Electronics of Solids. New York, McGraw-Hill 1965
4 Dekker, J.: Fyzika pevných látek. Praha, Academia 1966
5 Eckertová, L.: Fyzikální elektronika. Praha, Karlova universita 1970
6 Eckertová, L., Khol, F., Schröfel, J. (editor): Metody vytváření a kontroly tenkých vrstev. Praha, JČMF 1972
7 Elektronika tenkých vrstev. Praha, SNTL 1970
8 Ewing, J., Jardetsky, W., Press, J.: Elastic Waves in Layered Media. New York, McGraw-Hill 1967
9 Francombe, M., Sado, H. (editors): Single Crystal Films, New York, Pergamon Press 1964
10 Hasse, G. (editor): Physics of Thin Films. New York, Academic Press 1963−1967
11 Heavens, O. S.: Thin Film Physics. London, Methuen 1970
12 Holland, L. (editor): Thin Film Microelectronics. London, Chapman and Hall 1965
13 Chopra, K. L.: Thin Film Phenomena. New York, McGraw-Hill 1969
14 Keonjian, E.: Microelectronics. New York, McGraw-Hill 1963
15 Kittel, C.: Elementary Solid Physics. New York, John Wiley 1962
16 Khol, F., Pátý, L. (editors): Skripta letní školy tenkých vrstev. Praha, Socialistická akademie 1967
17 Lamb, D. R.: Electrical Conduction in Thin Insulating Films. London, Methuen 1967
18 Maissel, L. I., Glang, R. (editors): Handbook of Thin Film Technology. New York, McGraw-Hill 1970
19 Marton, L., El-Kareh, A. B. (editors): Electron Beam and Laser Beam Technology. New York, Academic Press 1968.
20 Mayer, H.: Physik dünner Schichten I, II. Berlin, Springer 1950, 1952
21 Methfessel, S.: Dünne Schichten. (Russian translation) Moscow 1963
22 Neugebauer, C. A., Newkirk, J. B., Vermilyea, D. A. (editors): Structure and Properties of Thin Films. New York, John Wiley 1962
23 Niedermayer, R., Mayer, H. (editors): Basic Problems in Thin Film Physics. Göttingen, Vandenhoeck - Ruprecht 1966
24 Paige, E. G. S.: Acoustic Surface Waves and Their Applications in Electronics, 7th Intern. Congress on Accoustics, Budapest 1971, p. 141

248

25 Pátý, L.: Fyzika nízkých tlaků. Praha, Academia 1968
26 Pashley, D.: Advances in Physics **14**, 1965, p. 327
27 Rutner, E., Goldfinger, P., Hirth, J. P. (editors): Condensation and Evaporation of Solids. New York, Gordon and Breack 1964
28 Sluckaja, V. V.: Tonkie plyonki i jich primenenie. Moscow, Gosenergoizdat 1962
29 Sodomka, L.: Struktura a vlastnosti pevných látek. Praha, SNTL 1967
30 Van Atta, C. M.: Vacuum Science and Technology. New York, McGraw-Hill 1965
31 Young, L.: Anodic Oxide Films. New York, Academic Press 1961
32 Pinsker, G.: Diffrakcia electronov. Moscow, GITTL 1953
33 Mayer, H.: Physik dünner Schichten, Gesamtbibliographie. Stuttgart, Wiss. Verlagsges. 1972
34 Groszkowski, J.: Technika vysokogo vakuuma, Mír, Moscow 1975
35 Kaminsky, M.: Atomic and Ionic Impact Phenomena on Metal Surfaces, Springer-Verlag, Berlin
36 Plešivcev, N. V.: Katodnoe Raspylenie, Atomizdat, Moscow 1968
37 Hartman, P. (editor): Crystal Growth, North-Holland Publ. 1973
38 Berry, R. W., Hale, D. M., Harris, M. T.: Thin Film Technology. Van Nostrand Co., Princeton, New Jersey 1970
39 Physics of Thin Films, Academic Press, from 1965
40 Bell, R. L.: Negative Electron Affinity Devices. Clarendon Press, Oxford 1973

Journals:
Thin Solid Films
Surface Science
Journal of Applied Physics
Solid-State-Electronics
Review of Scientific Instruments
Balzers-Informationen
Journal of Crystal Growth
Journal of Vacuum Science and Technology

# INDEX

accommodation coefficient 23, 73, 88
accumulation, surface layer 178
acoustic surface waves 160, 238
acoustic waveguide 161
adhesion 38, 98, 158
adsorption 18, 137, 242
   chemical 74, 97
   physical 97
ageing of films 211, 229
amorphous state 102, 196, 197
antireflection coating 226, 243
AT-cut 56

band
   bending 176, 242
   diagram 192, 199
   theory 162
barrier
   height 197, 200
   lowering 134
   transparency 197, 200
Bitter techniques 220
Boltzmann transport equation 165
Bragg condition 139
Bravais lattice 105
breakdown
   avalanche 212
   electrical 209
   thermal 212
   voltage 230
   Zener 199

cardiography 235
cathode dark space 20
cathode fall 19
cathode sputtering 18 ff
cavity resonator 235
cermet 229

channel electron multiplier 116
chemical transport reaction 16
chromatography 115
Clapeyron — Clausius equation 29
clean surfaces 42, 242
coalescence 72, 93
cold cathode 243
collision factor 86
condensation
   coefficient 59, 74, 87, 91
   process 85
conductivity, electrical
   of dielectrics 196
   of metal films
      continuous 163
      discontinuous 170
continuity equation 206
Cooper pair 235
corrosion 11, 225
   electrolytic 39
coupling element 239
crystal 104
   planes 106
   twins 109, 112, 143
crystallite, orientation 108
   size 147
crystallographic systems 104
Curie point 214

Debye length 176
Debye temperature 189
decoration 40, 41, 79, 101
depletion layer 178, 237
deposition, electrodeless 14
   chemical vapor 16
   electrolytic 14
   rate 52

desorption
  activation energy of   75, 90
  heat of   74
detectors   240
dielectric
  diode   208
  losses   210 ff
diffraction
  contrast   123
  electron   120, 122, 137, 139 ff
    high-energy   140
    low-energy   144
    reflection   143
    transmission   141
  scattering factor, atomic   139
    structural   139
  X-ray   141, 147
diffusion   206
  surface   74, 90
dislocation   108, 109, 156, 158
dissociation of pairs   84

eddy currents   53, 58
effective mass   177, 179
elastic properties   54
electrochemical equivalent   14
electroluminescent panel   244
electron beam   50, 120, 228
  evaporation   48
electron diffraction   139 ff
electron microscope   118 ff
  aberrations   121
  dark field   127
  emission   132
  interference   127
  Lorentz   220
  reflection   131
  scanning   130
  shadow   127
  transmission   118
electron scattering   122, 164, 176
  diffuse   166, 183
  specular   167
electro-transport   229
elementary cell   104
encephalography   235
enhancement operation mode   237
epitaxial film   238
epitaxy   16, 99, 110 ff

heteroepitaxy   16, 110
homoepitaxy   16, 110
liquid phase   17

rheotaxy   110
temperature   99, 111
etching
  high-frequency   50
  ion   50
evaporation   28 ff
  boats   45
  crucibles   45
  electron gun   47
  exploding wire   47
  flash   47
  getter   33
  heat
    external   29
    internal   28
  laser beam   47, 51
  materials   43
  multicomponent materials   32
  reactive   33
  sources   45
  three temperature methods   33

Fick's law   94
field effect   169, 179, 236
field electron microscope   91, 135
field ion microscope   138
filters   235, 239
  broad-band   226
  color   227
  grey   225
  ultranarrow band   227
Fowler-Nordheim equation   134
frequency stabilization   235
Fresnel relations   221
Fresnel coefficients   222
Fuchs-Sondheimer theory   165, 178, 182

GaAs   16, 17, 242
gaussian distribution   89
getter   24
  evaporation   33
  sputtering   24
Gibbs potential   77
Gibbs-Thomson equation   93
Gunn diode   235

Hall effect   179
heat radiation   225
hopping process   175

image forces   134, 171, 202
injection of carriers   206
integrated circuit   12, 227
  hybrid   240
  solid state   240
  thin film   240
integrated optics   244
interference   225
  colors   63
  filters   226
  fringes   64
  magnetometer   233
  maximum   62
  microscope   66
  minimum   62
  voltmeter   233
interferometer, Fabry-Perot   227
ionic conductivity   197
ionization gauge   59
islands   72, 170
  size distribution   95

Josephson cryotron   233
Josephson effect   193 ff
Josephson junction   195

Kikuchi patterns   142
Kohler's rule   181

large-scale integration   241
laser beam   51, 228
  evaporation by   47
lattice, crystal   104
  constant change   108
  constant difference   110
  defects   108
LEED   110, 144
lightguides   244
Lorentz force   180
loss angle   209 ff, 230
lubrication   224
luminous sensitivity   242
magnetic anisotropy   214
magnetic domains   169, 213
  walls   219

magnetic flux quantum   188, 194
  ripple   219
magnetization, spontaneous   213, 214
magnetometer   233, 235
magnetoresistance   181
  anisotropic effects   184
  longitudinal   182
  transverse   183
  Sondheimer theory of   183
Matthiessen rule   164
mean free path of electron   163, 172, 176
melting point of crystallites   96
memory element
  cryotron   233, 235
  ferromagnetic   232
metastable state   103
microbalance   91
microscopy
  electron   118 ff
  X-ray   147
microwave generation   235, 236
migration   74, 137
Miller indices   106
mirrors   225
  dichromatic   226
misfit   111, 156
mixing   236, 240
mobility   172, 180, 206, 208
modulator   240
moiré pattern   108, 109, 124
molecular field model   214
monocrystalline film   102, 110

negative resistance   193, 207
nucleation
  heterogeneous   78
  homogeneous   77, 80
  rate   79, 87
nucleus, critical   78, 81, 82, 84, 90

ohmic contact   200, 206
oscillatory circuit   235
oxidation
  anodic   15, 230
  plasma   15
  thermal   15
passivation of silicon   16
photocathode
  cesium-antimony   242

multialkali 242
negative electron affinity 242
photolithography 49, 228
photolysis 16
Poisson equation 206
polarization
  dielectric 210
  optical 221
polarography 115
polycrystalline film 102
Poole-Frenkel effect 199, 204
pre-nucleation 89
primitive cell 104
pseudomorphism 108
pyrolysis 16

Q-factor 235
quadrupole mass filter 116
quantum size effects 162, 179

radiation detectors 236, 242, 243
recrystallization 97
rectifying effect 10, 208
reflectance 60, 64, 227
reflectors astronomical 225
reflex klystron 235
refractive index 61, 221
relaxation processes 210
relaxation time 166, 176
replica 129
resistivity, of metal films
  continuous 163, 228
  discontinuous 171
resistivity temperature coefficient 168
resonator 239
rheotaxy 110
Richardson equation 171
Richardson-Schottky equation 204

sandwich structure 195, 196 ff
Schottky effect 171, 172, 199, 204
secondary emission 243
separate media amplifier 239
SIMS 116
size effects 165, 176
  galvanomagnetic 179, 182
  quantum 172, 179
skin effect 58
space charge limited currents 200, 205

spectral analysis 115
spectroscopy, Auger 145, 148
  mass 92, 115
  secondary ion mass 116
spin wave model 214
sputtering, cathode 18 ff
  coefficient 20
  diode 18
  getter 24
  high-frequency 25, 27
  reactive 25
  threshold energy 21
  triode 25, 26
statistical theory of nucleation
  Rhodin-Walton 81
  Zinsmeister 83
sticking coefficient 73
Stoner-Wohlfarth asteroid 217
strain-stress dependence 153, 157
strength measurement 154, 155
stress, internal
  intrinsic 155, 156
  measurement 152
  thermal 155
substrates
  ceramic 40
  cleaved single crystals 40
  glass 39
sublimation 28, 37
sun-glasses 227
superconductivity 184 ff
  Bardeen-Cooper-Schrieffer theory 187
  coherence range 187
  Cooper pairs 187, 193, 235, 236
  critical magnetic field 185, 191
  critical temperature 184
  hard superconductor 187
  Josephson junction 195, 235
  London penetration depth 185
  Pippard-Ginzburg-Landau theory 186
  quantum generator 195
  soft superconductors 187
  SQUID 235
  weak coupling 193
supersaturation 76, 77
  critical 80
superstructures 110, 146
surface diffusion 74
  activation energy 90, 112

coefficient   75, 94, 137
surface tension   225
switching circuit   241

texture   108, 141
  fiber   108, 109
  fiber texture axis   109
thickness,
  mass   52
  measurement   52 ff
thin film transistor   236
time of residence   74, 84
Tolansky method   65
transducer
  compression   161
  interdigital   161, 239
transmittance   221, 225
Trouton relation   81
tunneling   134, 171, 172, 199, 200
  activated   173
  between superconductors   191
  of Cooper pairs   193

vacuum pump
  cryogenic   37
  ion-getter   35
  oil diffusion   34
  sublimation   37
  turbomolecular   37
van der Waals forces   90

waveguides   239
weak-absorbing substances   227
Wheatstone bridge   67
whiskers   154
WKB-method   201
work function   71, 136, 171, 242

X-ray, diffraction   147, 153
X-ray fluorescence   115
X-ray microanalysis   115
X-ray microscopy   147

Zeldovich constant   79
Zener breakdown   199